CALIFORNIA NATURAL HISTORY GUIDES

INTRODUCTION TO CALIFORNIA BIRDLIFE

California Natural History Guides

Phyllis M. Faber and Bruce M. Pavlik, General Editors

Introduction to CALIFORNIA BIRDLIFE

Text by Jules Evens
Photography by Ian Tait

UNIVERSITY OF CALIFORNIA PRESS
Berkeley Los Angeles London

This book is dedicated to Rich Stallcup and Bob Stewart in appreciation of their wide-ranging knowledge of the natural world, their generous teaching, and their genuine love of native birds.

California Natural History Guide Series No. 83

University of California Press
Berkeley and Los Angeles, California

University of California Press, Ltd.
London, England

Library of Congress Cataloging-in-Publication Data

Evens, Jules G.
Introduction to California birdlife / text by Jules Evens ; photography by Ian Tait.
p. cm.
Includes bibliographical references and index.
ISBN 0-520-23861-3 (cloth : alk. paper)—ISBN 0-520-24254-8 (pbk. : alk. paper)
1. Birds—California. I. Title.

QL684.C2E94 2005
598′.09794—dc22 2004013038

Manufactured in China
10 09 08 07 06 05
10 9 8 7 6 5 4 3 2 1

The paper used in this publication meets the minimum requirements of ANSI/NISO Z39.48–1992 (R 1997) (*Permanence of Paper*). ♾

Cover photograph: Yellow-billed Magpie *(Pica nuttalli)*. Photograph by Ian Tait.

The publisher gratefully acknowledges the generous contributions to this book provided by

the Moore Family Foundation
Richard & Rhoda Goldman Fund
and
the General Endowment Fund of the
University of California Press Associates.

CONTENTS

One touch of nature makes the whole world kin; and it is truly wonderful how love-telling the small voices of the birds are, and how far they reach through the woods into one another's hearts and into ours.

JOHN MUIR, 1854

If a parent wishes to give his children three gifts for the years to come, I should put next to a passion for truth and a sense of humor, love of beauty in any form. Who will deny that birds are a conspicuous manifestation of beauty in nature?

RALPH HOFFMANN, 1927

Most of us who follow birds are partly scientist, partly sportsman, and partly poet.

LEON DAWSON, 1923

PREFACE

What is a bird? This might seem a simple question, easily answered. "A warm-blooded, egg-laying, vertebrate animal with feathers" would be accurate, but would only begin to get at the truth. It could be added that the forelimbs are modified to form wings and these are used for flight; this would also be true. Birds are therefore bipedal, using their two hind limbs for walking, running, swimming, perching, foraging, and sometimes killing prey. Also, birds have bony protrusions covered in keratin that surround the mouth—called beaks, or bills. Oh yes, and they are toothless. But here the shared characteristics that define a bird end.

Evolution has found in the class Aves one of the most pliable and adaptable of its creations. Perhaps more than any of the other vertebrates—fish, amphibians, reptiles, or mammals—birds are highly responsive to the challenges, changes, and opportunities of the environments in which they live. As a result, the class has diverged from a common ancestor (probably a theropod dinosaur) into a marvelous multiplicity of species (currently about 10,000 known species living on Earth), each with its own unique attributes and abilities. Each species is an eloquent expression of the environment in which it lives, and most native species fit seamlessly into their own habitat niches. The Northern Spotted Owl is a manifestation of the moist, shady forests of the North Coast range. The California Black Rail is as much a part of the tidal marsh as pickleweed or marsh rosemary. The Greater Sage-Grouse is a creation of the sagebrush plains of the Great Basin.

To know California's birds, one must know the habitat in which they live. This affinity between birds and their habitats is the theme of this book. Only through an understanding of those places—of their weather patterns, of their plant communities, of their essential natures—will one discover the nature of the most lively residents, the birds. This book is organized according to California's bioregions, each of which supports its own complement of bird species. We have attempted to provide insight into the ecological dynamics of each bioregion and into the distribution, occurrence, and behavioral adaptations of its representative birds.

This is not a field guide. There are a plethora of those available, and you may want to keep one handy to look up species that are not illustrated by a photograph. Nor is this an ornithology textbook. Those are also readily available; we suggest the National Audubon Society's *The Sibley Guide to Bird Life and Behavior* (2001) for encyclopedic access to general ornithological information.

This is a book about California's birds. California is such a diverse and vibrant environment, and supports such bountiful birdlife, that we have been able to paint only a partial picture of its ornithological richness. But it is our hope that this introduction will set you off on a lifelong journey of exploration and discovery in the company of California's birds.

ACKNOWLEDGMENTS

The libraries at Point Reyes Bird Observatory, Audubon Canyon Ranch's Cypress Grove Preserve, and the library at the University of California at Berkeley Life Sciences building provided essential source material for this book. Earlier versions of the text were greatly improved by the editorial comments of Jeff Davis, Phyllis Faber, Steve Howell, Doris Kretschmer, Scott Norton, Ian Tait, and Steve Laymon. Specific information and field knowledge was generously provided through discussions or communications with David DeSante, Meryl Evens, Alan Fish, Sam Fitton, Mary Anne Flett, Keith Hanson, Phil Henderson, Steve Howell, John Kelly, Jack Nisbet, Gary Page, Claire Peaslee, Peter Pyle, Ane Carla Rovetta, Dave Shuford, Rich Stallcup, Lynne Stenzel, Bob Stewart, Emilie Strauss, Philip Unitt, David Wimpfheimer, Jon Winter, and many other unnamed companions, biologists, and birders encountered in the field. Thank you all. — J. E.

Photographing birds is mostly a solitary occupation. However, successful results often depend on the assistance of friends and ornithological professionals. Among these, I would especially like to thank Walter Koenig, Steve Laymon, Bob Stewart, Pam Williams, and several generations of Point Reyes Bird Observatory biologists. — I. T.

AN OVERVIEW OF CALIFORNIA BIRDLIFE

California is a bonanza of birdlife. The more than 600 bird species on the California state list represent about three-quarters of the 800 or so species that have been recorded in continental United States and about two-thirds of the more than 900 species that occur in North America north of Mexico. Nearly half (47 percent) of California's bird species breed in the state; the rest come to spend the winter in the hospitable climate, or pass through migrating to other wintering or breeding grounds. A relatively large number, perhaps 25 percent of birds on the California list, are rare in the state, occurring only occasionally as *vagrants,* when anomalous weather patterns, ocean currents, or even pioneering tendencies lead them into the state.

California lies at an intersection of atmospheric and oceanic currents and, therefore, at the crossroads of migratory bird routes of the Western Hemisphere. Within a single day, an enthusiastic California birder, following a well-planned schedule and with attention to local knowledge and weather conditions, can encounter a wide variety of birds from far-flung regions of the globe—flocks of Sooty Shearwaters *(Puffinus griseus)* from the South Pacific, Elegant Terns *(Sterna elegans)* from Northwest Mexico, Mountain Plovers *(Charadrius montanus)* from the Colorado Plateau, Black-bellied Plovers *(Pluvialis squatarola)* from the Alaskan tundra, a Rough-legged Hawk *(Buteo lagopus)* visiting from the Canadian prairie, and even an out-of-range warbler, maybe an American Redstart *(Setophaga ruticilla)* or an Ovenbird *(Seiurus aurocapillus)* from New England's hardwood forests. On that same day, with some careful searching through a variety of habitats, the curious naturalist is sure to see dozens, or perhaps more than a hundred, species of birds common throughout much of the west, and several species that occur almost exclusively in California.

This exceptional diversity of birds exists because of California's relatively equitable climate and a varied topography that supports a crazy quilt of habitats. The state's ecological heterogeneity—from dry desert washes, to vast valley grasslands and wetlands, to mossy coastal rainforests and fertile estuaries—provides refuge and sustenance for as rich a complement of bird species as can be found anywhere in North America. Like the geology, the weather, the flora, and the human population, the avifauna of California is constantly changing, adding another layer of natural vitality and vibrancy to this exuberant landscape.

The number of bird species recorded in California keeps increasing, gradually but inexorably—an example of the overall dynamism that is so characteristic of the Golden State. This expansive trend is gradually fulfilling the prophecy of the grandfather of California ornithology, Joseph Grinnell (1877–1939): "It is only a matter of time theoretically until the list of California birds will be identical with that for North America as a whole."

The state's vast size and its Pleistocene climatic history contribute to a high degree of *endemism*. Indeed, California is the only mainland portion of the United States recognized as an Endemic Bird Area, because of its several endemic species and many endemic subspecies. (See "Endemics and Near Endemics," below.)

Terms

Specific terms related to avian biology are discussed later in this overview (e.g., "Taxonomy," below) or are italicized with definitions provided in the glossary. Some of the general terms used to describe types of birds are not very precise, but rather refer to general behavioral characteristics. In broadest terms, the phrases "waterbirds" and "landbirds" divide the class Aves into those species whose primary habitat is aquatic and those whose primary habitat is terrestrial. Most field guides are organized taxonomically, with waterbirds (loons through alcids) in the first half and landbirds (doves through finches) occupying the second half of the guide. This order is based on our understanding of the evolutionary sequence of avian families, from the most primitive to the most recent. A few bird families do not fit neatly into these large groupings, however. Terrestrial groups such as the diurnal birds of prey *(raptors)* and upland game birds (quail and grouse) are inserted between true waterbirds such as ducks and sandpipers.

Within each of these major divisions are other distinctions that help to describe physical similarities or habitat preferences among groups of species. Waterbirds may be "seabirds" or "waterfowl," or "waders" or "shorebirds." Seabirds include a wide variety of families and species with diverse taxonomic relationships, from shearwaters and albatrosses to murrelets and puffins, all of which occur in offshore, oceanic waters. "Waterfowl" is a term more limited taxonomically, referring to members of the family

Anatidae—swans, geese, and ducks—most of which are associated with interior or coastal wetlands. Some cross-pollination occurs among terms, however. For example, the Brant (*Branta bernicla,* also called Sea Goose) or the Surf Scoter (*Melanitta perspicillata,* a diving duck) may also be considered seabirds, because each is often found in oceanic waters. Further subdivisions are used to described subsets within these broader catchall categories. Ducks, for example, are described as either "diving" ducks or "dabbling" ducks, depending on their feeding behavior. The use of the term "waders" varies among people and cultures, but in the United States, it is generally used to describe long-legged wading birds such as herons and egrets. (In Europe, "waders" refers to shorebirds.) Shorebirds, many of which are generically called "sandpipers," are included in the diverse order Charadriiformes, which also includes the jaegers, gulls, and terns.

"Landbirds" is also a catchall phrase. In broadest terms it includes doves through finches. Some groups of birds are unique, and thus they are readily recognizable as owls or woodpeckers or hummingbirds. Others, such as vireos (Vireonidae) and warblers (Parulidae), or sparrows (Emberizidae) and finches, are much more similar to one another, and distinctions emerge only through familiarity, the result of careful observation. You will find it helpful to keep one of the popular field guides handy while reading through this book, to look up those species mentioned but not illustrated in the text. Familiarity with the organization and contents of the field guide will help you develop identification skills and learn taxonomic relationships; but nothing is more instructive than experiencing birds in real life.

Bioregions

We have organized California into seven bioregions to discuss the avifauna: Marine Environment, Shoreline, Coast Ranges, Central Valley and Delta, Mountains and Foothills, Great Basin, and Deserts. Other books in this natural history series organize the state's bioregions differently, depending on their subjects (e.g., Schoenherr 1992, Manolis 2003). Because birds are so mobile, and because most species are quite widely distributed, we have lumped several regions that could have been treated separately,

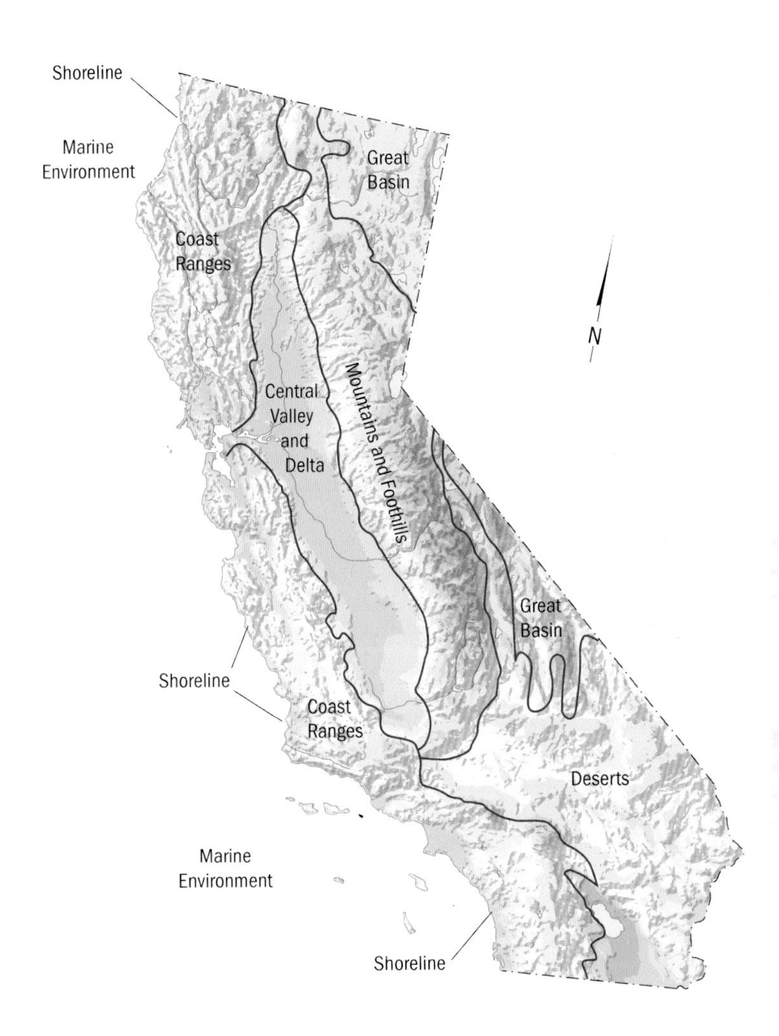

Map 1. Major bioregions of California.

and have split others. The Mojave and Colorado Deserts are lumped into "The Desert's Birds," the Great Basin and Modoc Plateau into "Birds of the Great Basin." The Ocean is treated separately from the Shoreline, which is treated separately from the Coast Ranges. "Birds of the Coast Ranges" includes the Klamath Mountains as well as the Transverse and Peninsular Ranges. Each of these regions could be subdivided into smaller units, and the observant naturalist will begin to make distinctions within each broad bioregion soon after exploring any given area.

The habitat boundaries and overlaps within bioregions and the ways in which different species and communities of birds use the resources of each habitat is a major focus of this book. Birds are among the most adaptable animals, however, and many species occur across several bioregions, climatic zones, or plant communities, their distribution often varying with season and resource availability. In general, birds are highly responsive to changing environmental conditions and adept at exploiting new opportunities. For example, some species—Tricolored Blackbird *(Agelaius tricolor)*, Red Crossbill *(Loxia curvirostra)*, and Lawrence's Goldfinch *(Carduelis lawrencei)*—may nest in certain areas in some years and be absent the next. Some montane-breeding species—such as Red-breasted Nuthatch *(Sitta canadensis)* and Golden-crowned Kinglet *(Regulus satrapa)*—are

Plate 1. Golden-crowned Kinglets favor the canopy of coniferous forest. In some years, particularly during severe winters, irruptions of Kinglets appear in the lowlands, especially along the coast and in riparian areas of the Central Valley.

Plate 2. An American Robin brings a large beakful of earthworms to feed its hungry young. Robins have responded well to human settlement in California, expanding their range into residential areas.

irruptive, invading lowlands and Coast Ranges when winter weather is severe in the high mountains, but may be largely absent from the lowlands during milder winters.

No species is so broadly distributed as to occur throughout all bioregions, but a few are found in nearly every California habitat. Those species that have proved adept at adapting to human settlement and altered environments—such as American Robin *(Turdus migratorius)*, Brown-headed Cowbird *(Molothrus ater)*, and Brewer's Blackbird *(Euphagus cyanocephalus)*—may be found in almost any terrestrial habitat or bioregion except the highest alpine fell or the most desolate desert floor.

The boundaries of each bioregion are discussed in a general way in each corresponding chapter of this book. Some of these boundaries are admittedly arbitrary, and an argument to expand or reduce any given area can be made. Adjacent regions have overlapping boundaries, where they share species and communities. An obvious example is found in the Great Basin. The upper edge of the sagebrush steppe, the characteristic plant community of the Basin, grades into a pinyon–juniper forest, and sagebrush-dependent birds such as Greater Sage-Grouse *(Centrocercus urophasianus)* and Sage Thrasher *(Oreoscoptes montanus)*, range up into this zone, particularly where the trees are spaced widely and the understory is shrubby. The pinyon–juniper zone is considered part of the montane region here, but it might also have been covered in "Birds of the Great Basin." Similar examples occur at habitat edges between the Coast Ranges and the Central Valley, or the Sierra foothills and the Central Valley, or the Peninsular Ranges and the desert.

Many of California's counties are as large and diverse as some states, and many birders pay particular attention to the birdlife of a given county, maintaining county bird lists as an ongoing avocation, documenting overall biodiversity, and hoping to discover breeding species or vagrants never before recorded in that county. The coastal counties are the most hospitable to birds (and people); moderated by moist marine air and with an equitable coastal climate and a variety of habitats, they offer a wide range of *niches* for a wide variety of species. San Diego County, with 488 bird species at last count (including seven exotic species), hosts the greatest variety, not only in California but in any county in the United States. The coastal counties of Los Angeles, Monterey, and Marin follow closely behind, each with more than 480 species recorded. Interior counties tend to host fewer species, but each has its own unique avifauna and each has the potential for unexpected occurrences and new discoveries. Imperial County includes the northwestern reaches of the Sonoran Desert and so attracts several species whose ranges barely extend into California from Mexico and Arizona, such as the diminutive, cactus-dwelling Elf Owl *(Micrathene whitneyi)*, the sparrowlike Abert's Towhee *(Pipilo aberti)*, and the brassy Bronzed Cowbird *(Molothrus aeneus)*. (Elf Owl, in fact, may now have been *extirpated* from the state.) The Salton Sea, also in Imperial County, is in effect a northern extension of the Gulf of California and attracts a surprising array of wayward seabirds. Mono and Modoc Counties are western extensions of the Great Basin and account for the incursion of many Great Basin species into California.

Unique topographic or geomorphic features may attract species that would otherwise be absent from a given area. Such hot spots tend to be located at the boundaries of bioregions where several habitat types overlap. Kern River Preserve, in the southern Sierra near the western edge of the Transverse Ranges and the upper San Joaquin Valley, is one such crossroad. Morongo Valley, at the base of the San Bernardino Mountains and the head of the

Coachella Valley, where the Mojave and the Colorado Deserts merge, is another nexus of avian activity and overlapping distributions. Unique geomorphic features such as the White-Inyo, San Bernardino, New York, Clark, and Warner Mountains and Death Valley each have their own unique avian dynamics. Large interior wetlands such as the Salton Sea, Mono Lake, and the Klamath Basin are bird magnets, attracting huge numbers of birds. Famous coastal bird hot spots, where pelagic and landbird vagrants regularly appear, include Point Loma (San Diego County), Point Mugu (Ventura County), the Big Sur and Carmel River mouths (Monterey County), the Point Reyes Peninsula (Marin County), the Farallon Islands (San Francisco County), and the Smith River mouth (Del Norte County), to name a few.

Hot spots are not always grand mountain ranges, dramatic coastal headlands, or vast lakes. In arid lands, very small water sources, strategically located oases, also attract transients, especially migrant songbirds, and swell the county list out of proportion to the broad habitat type of the area; classic examples are Butterbredt Springs (Kern County), Soto Ranch (San Bernardino County), and Furnace Creek Ranch, Mesquite Springs, and Scotty's Castle in Death Valley (Inyo County).

The birds most familiar to Californians are the widely distributed species, or those most adapted to urban and suburban landscapes. Some nearly ubiquitous native species are soon recognized by the observant beginning naturalist—a Red-tailed Hawk *(Buteo jamaicensis)* soaring on broad wings over open fields or rolling hills; that brilliant aerobat, Anna's Hummingbird *(Calypte anna)*, buzzing busily around gardens and feeders; the garrulous Western Scrub-Jay *(Aphelocoma californica)*, its back as blue as the southern California sky. Other birds are as familiar in California as in the rest of the country: the soft-cooing Mourning Dove *(Zenaida macroura)*; the Northern Flicker *(Colaptes auratus)* (except in the deserts); American Robin, American Goldfinch *(Carduelis tristis)*, House Finch *(Carpodacus mexicanus)*, Red-winged Blackbird *(Agelaius phoeniceus)* and Brewer's Blackbird, and of course Common Raven *(Corvus corax)* and American Crow *(C. brachyrhynchos)*. Several nonnative, or introduced, species are also as common here as elsewhere: Rock Pigeon (or "Rock Dove") *(Columba livia)*, European Starling *(Sturnus vulgaris)*, and the "English" House Sparrow *(Passer domesticus)*.

Birds with strict habitat associations, relatively limited or

Plate 3. The male House Finch courts the female by singing, wing fluttering, and courtship feeding. House finches are cosmopolitan, found in many terrestrial habitats from mountains to deserts.

patchy distributions, or secretive habits are less familiar. Great Gray Owl *(Strix nebulosa)* is a reclusive resident of high mountain meadows, and Flammulated Owl *(Otus flammeolus)* is a common but highly nocturnal (and therefore rarely seen) breeder in mixed conifer forests. Black-backed Woodpecker *(Picoides arcticus)* is a quiet denizen of the lodgepole pine *(Pinus contorta* subsp. *murrayana)* belt and mountain forest burns. Black Swift *(Cypseloides niger)* is a rather rare aerial insectivore that forages at great heights and is glimpsed only fleetingly, darting into nesting sites behind mountain waterfalls! Species such as Black Rail *(Laterallus jamaicensis)*, Least Bittern *(Ixobrychus exilis)*, and Yellow-breasted Chat *(Icteria virens)* are so inconspicuous and furtive that most are detected only by sound, or not at all. The very restricted distribution of Gray-crowned Rosy-Finches *(Leucosticte tephrocotis)* in high alpine snow fields means that they rarely see people, and vice versa. So, too, with the whole community of *pelagic* species; millions of seabirds spend their lives on offshore islands and where oceanic currents converge, well beyond the boundary of normal human experience.

Birds strictly associated with specific plant communities or habitats are rarely encountered elsewhere. California Clapper Rail *(Rallus longirostris obsoletus)*, so chickenlike it was formerly called the "marsh hen," leaves the tidal marsh only during extreme flood tides; Pinyon Jay *(Gymnorhinus cyanocephalus)* seldom wanders beyond the pinyon–juniper forest because of its diet of pinyon nuts. Nondescript seedeaters of California's native shrublands are also homebodies. Bell's Sage Sparrow *(Ampispiza belli belli)* is rarely found outside of chamise *(Adenostoma fasciculatum)* shrub habitat of the dry, "hard" chaparral, and Rufous-crowned Sparrow *(Aimophila ruficeps)* has a strong preference for coastal sagebrush *(Artemisia californica)* in damper "soft" chaparral habitats such as occurs along the coast.

Sometimes habitat affinities are very strict during the breeding season and broaden during the nonbreeding months. For example, Sage Thrashers are confined to the sagebrush belt of eastern California as breeders, but visit the coastal slopes of southern California in winter. Many high-mountain breeders move downslope to lower elevations in winter, and, conversely, low-elevation breeders drift upslope after breeding. Yet other species such as Pine Siskin *(Carduelis pinus)*, Red Crossbill, and Lawrence's Goldfinch, nomads that wander widely in search of variably abundant seeds, are more erratic in their distribution.

Habitat

Range and Habitat

The terms "range" and "habitat" are often used in reference to the distribution of birds, and it is important that the two concepts not be confused. The range of a species is an outline of the area in which the species normally occurs. Within that broad range are patches of appropriate habitat, marginal habitat, and unsuitable habitat. For example, a range map of Red Crossbill might show it occurring all along California's north coast, but Crossbills eat seeds of coniferous trees and therefore regularly occur only in habitat that provides that critical resource. Rarely, if ever, are Crossbills encountered in coastal prairie, chaparral, or oak woodland habitats.

Critical Habitat Components

In subsequent chapters, we have taken a bioregional-habitat approach to California birdlife, discussing the common bird communities within regional habitat types. Some bird species have general habitat affinities, but rely on some specific habitat components within that community for sustenance. The Phainopepla *(Phainopepla nitens)*-mistletoe *(Phoradendron flavescens)* connection is an obvious example. Phainopepla occurs in desert wash and valley and foothill woodlands, but within those habitats it is closely associated with mistletoe infestations—whether on oak or mesquite. There are many examples of such close associations: Hermit Thrush *(Catharus guttatus)* and coffeeberry (*Rhamnus* spp.); Lawrence's Goldfinch and fiddlenecks (*Amsinckia* spp.); White-tailed Kite *(Elanus leucurus)* and meadow voles (*Microtus* spp.); Townsend's Solitaire *(Myadestes townsendi)* and juniper (*Juniperus* spp.) berries; Scott's Oriole *(Icterus parisorum)* and Joshua trees *(Yucca brevifolia)*; Hooded Oriole *(Icterus cucullatus)* and fan palms *(Washingtonia filifera)*. Some relationships are so critical to the species' life history that the relationship is obligatory and has been incorporated into that species' common name:

Plate 4. The male Phainopepla is often seen perched on a treetop, especially in areas where mistletoe berries are abundant. His black plumage distinguishes him from the gray-colored female, an example of sexual dimorphism or dichromatism.

Plate 5. Spotted Towhees are found regularly in all bioregions except deserts. They build their well-hidden cup-shaped nests on the ground and in shrubs in dense thickets.

Greater Sage-Grouse, Acorn Woodpecker *(Melanerpes formicivorus)*, Willow Flycatcher *(Empidonax traillii)*, Sage Thrasher, Oak and Juniper Titmouse *(Baeolophus inornatus* and *B. griseus)*, Marsh Wren *(Cistothorus palustris)*, and Sage Sparrow *(Amphispiza belli)* are obvious examples. But some common names mischaracterize habitat associations; surely Barn Owls *(Tyto alba)* were around before there were barns and Roadrunners *(Geococcyx californianus)* before there were roads! No matter how strict these alliances seem, some subtleties to the relationship usually become apparent only after close scrutiny. So, we find that Yellow Warbler *(Dendroica petechia)* breeding success depends not only on the presence of extensive willow thickets, but also requires that youthful willows should be growing near more mature stands. Similarly, California Clapper Rails occupy tidal marshes where not only older stands of cordgrass (*Spartina* spp.) occur but also newer marsh is actively developing. Spotted Owls *(Strix occidentalis)* need not only old-growth forest with a closed canopy but also a complement of decadent trees and decaying woody debris that supports abundant fungal growth, and thus a healthy rodent population, which thrives on the fruiting bodies of the fungi. Birds are products of their habitats, and, conversely, the habitat's

Plate 6. A night-hunting Barn Owl brings food for its nestlings hidden in the grain hopper of an abandoned mill. Barn Owls can be found in open woodlands and grasslands throughout the state.

health is dependent on the vitality of species interactions within it. Nature's elegance is matched only by her complexity, and that complexity is understood only through careful, and often long-term, observation.

Evolution has provided birds with subtle variations in morphology and behavior that allow similar species to forage side by side while exploiting different resources from the habitat and avoiding competition. Thus, a flooded vernal pool the size of your living room may host several dabbling ducks: Cinnamon Teal *(Anas cyanoptera)* siphon microalgae from the water's surface, whereas Green-winged Teal *(A. crecca)* slurp invertebrates from the muddy edge; Mallard *(A. platyrhynchos)* graze on roots and rhizomes, and Northern Pintail tip-up to pick copepods and midge larvae from the pool's bottom.

Edges and Ecotones

It is tempting, and perhaps unavoidable, for the human mind to separate habitats into discrete units and their inhabitants into discrete communities. Such natural organization does exist, but in such dynamic variety that even the most thorough description is confounded by the complexity and plasticity of boundaries among and between various habitats. Add to this the flexibility of natural systems and the extraordinary adaptability of birds, and the confusion is amplified. This dynamism, this creativity, is what makes nature study and bird study so engaging. Most "habitats" are really a mosaic of habitat elements woven into a tapestry of interacting relationships. It is instructive to recognize that across North America there is a higher degree of natural habitat fragmentation in the west, particularly in California, than in the east, because of greater aridity, greater frequency of wildfires, and greater topographic diversity. Because of this natural heterogeneity, habitat edges are a common feature of the western landscape.

Habitats have cores and edges, and the relationship between them is an important facet of biodiversity. The ratio of edge to core depends on the size of the habitat patch; smaller patches of habitat have more edge relative to core and larger patches have less. Birds, like many other organisms, respond to edges, areas of overlap between habitats. These zones of habitat interspersion, these *ecotones,* may contain a greater number of species and higher population densities of some species than either adjoining community. Many bird species, "edge exploiters," are attracted to these habitat boundaries or occur only where several habitat types intersect; other species, "edge avoiders," avoid edges. A study of edge sensitivity in coastal southern California found that Anna's Hummingbird, Northern Mockingbird *(Mimus polyglottos),* Western Scrub-Jay, and House Finch occur in higher numbers at the edge than in the core of a habitat (Bolger 2002); conversely, Sage Sparrow, Rufous-crowned Sparrow, and Black-chinned Sparrow *(Spizella atrogularis)* are fewer where the edge is greater. Predation and nest parasitism may be higher at the forest edge than in the deep forest, where more cover is available (see sidebar "The Brown-headed Cowbird"); a high rate of nest predation has also been found to be a function of habitat patch size, however.

Given California's topographic, climatic, and ecological diversity, edge habitat, or overlapping habitat mosaics, may be more extensive than habitat cores, areas of homogeneous vegetation types. Furthermore, ecotones and edges have been enlarged by human alteration of the landscape, especially since European arrival. (Some edges, especially those internal to habitat types, have decreased in extent, however, with conversion of multi-

Plate 7. The ragged, bag-shaped nest of a Bushtit is usually built in a tree, camouflaged by hanging twigs and branches. When not attending a nest, Bushtits are usually found in loose groups, actively gleaning insects from tree foliage.

structural, multispecies habitats to monocultures. For example, complex coastal prairie or sagebrush steppe communities have been converted to relatively homogeneous annual grasslands.) Edges are focal points for many species, and those species have responded to the increase in edge that has accompanied the transformation of California. Many of the species most common in residential and suburban settings are in fact "edge specialists." Residential areas and population centers such as the San Francisco Bay Area, Santa Barbara, and San Diego have been converted from coastal scrub habitats to lush oases enhanced by irrigation and ornamental plantings to the benefit of many of the birds most familiar to Californians—Anna's Hummingbird, Western Scrub-Jay, Bushtit *(Psaltriparus minimus)*, American Robin, House Sparrow, Brewer's Blackbird, and Northern Mockingbird. These generalists have increased in abundance in response to habitat conversion. Other, more sensitive or specialized species have decreased accordingly. Coastal sage scrub habitats of the southern coast have been among the most heavily impacted habitats, and the numbers of California Quail *(Callipepla californica)*, California Gnatcatchers *(Polioptila californica)*, and Greater Roadrunners have suffered. Tidal marshes have been all but obliterated outside of (and substantially reduced within) San Francisco Bay, with the inevitable loss of Clapper Rails *(Rallus longirostris)* and Black Rails, Song Sparrows *(Melospiza melodia)* and Savannah Sparrows *(Passerculus sandwichensis)*, and Common Yellowthroats *(Geothlypis trichas)*, among others. Riparian corridors have been lost almost entirely in the Central Valley, a tragedy that decisively extirpated Warbling and Bell's Vireos *(Vireo gilvus* and *V. bellii)*, Yellow-billed Cuckoo *(Coccyzus americanus)*, and other riparian loving birds.

Those species that have evolved in response to the unpredictability and variability of the edge habitats may persevere, depending on how strict their habitat association is. A certain degree of serendipity comes into play, with habitat availability influenced by fire, drought, logging, agricultural practices, and other vagaries of nature and man. Lawrence's Goldfinch, a seedeating *fringillid* nearly endemic to California and Baja, tends to occur where three habitat types overlap—dry and open oak

(continued on page 20)

Plate 8. Northern Flickers normally excavate a new nest hole every breeding season but occasionally use a hole from a previous year. These distinctively large woodpeckers breed at forest edges from the coast to the mountains.

THE BROWN-HEADED COWBIRD

Perhaps no species is as peripatetic as the Brown-headed Cowbird *(Molothrus ater),* which invaded nearly every corner of the state in the twentieth century. Since 1900, the Cowbird has gone from being an irregular winter visitor of the southeastern corner of the state to one of the most wide-ranging and abundant members of the California avifauna, and also, the most detrimental. Cowbirds are "brood parasites"; they lay eggs in the nests of other species and abandon them to the care of the "hosts," which become foster parents. This behavior relieves the Cowbird of those behaviors that are so expensive, in terms of energy, and leaves them to other species—territory defense, nest building, incubation, and chick rearing. This lack of responsibility allows Cowbirds to wander widely and devote all of their breeding effort to finding host nests in which to deposit their eggs. Female Cowbirds are unusually prolific; a single female can lay 30 to 40 eggs per season. As Elliot Coues observed (1903), the Cowbird shows "a remarkable exception to the rule of conjugal affection and fidelity among birds . . . [and] a wonderful provision for perpetuation of the species is seen in its instinctive selection of smaller birds as the foster-parents of its offspring." Many of the host species most susceptible to Cowbird parasitism are also those species—mostly small passerines—that have declined so dramatically over the last century as the Cowbird has increased.

Plate 9. A Brown-headed Cowbird performs his courtship display to attract females.

What has caused this increase? Before European settlement of North America the Cowbird was a bare-ground specialist, a bird of the shortgrass prairie of Great Plains associated with hoofed mammals, such as bison and antelope. According to Coues (1874), "every wagon train passing over the prairies in summer was accompanied by cowbird flocks" (quoted in Rothstein 1994). After the wagon trains arrived and settlements were established, the new settlers began clearing land and converting the landscape to agricultural endeavors, in effect creating Cowbird habitat. Rothstein (1994) traces the Cowbird's colonization of California. Although absent earlier, Cowbirds were first discovered in the San Joaquin Valley in 1907 when eggs were found in the nest of a Bell's Vireo. By the early 1930s Cowbirds had expanded into the northern reaches of the Central Valley, the west slope of the Sierra, and the Bay Area. The Tahoe Basin was not invaded until the mid-1950s, and by then the higher elevations of the Sierra were being visited by Cowbirds, probably related to intensive logging; Cowbirds seem to avoid dense forests, which explains, perhaps, why they were absent from northwestern California until the mid-1940s, when logging activity opened up the north coast mountains.

Plate 10. The size difference between the young Brown-headed Cowbird (on right) and the nestlings of the Wilson's Warbler host becomes apparent a few days after hatching.

The spread of Cowbirds throughout California, then, coincided with clearing of the native vegetation, the nesting "substrate" of many small songbirds. As discussed later in this book, reduction of riparian habitat was particularly severe in the twentieth century, caused by either intentional clearing for agricultural crops or inadvertent clearing by cattle, which trample and graze riparian foliage, especially young willows, and so degrade the habitat for nesting birds. The reduction in nesting cover, in turn, increased the opportunity for Cowbirds to discover and parasitize nests. Willow-nesting species—Willow Flycatcher *(Empidonax traillii)*, Bell's Vireo, and Yellow Warbler, in particular—have been particularly hard hit by the dual impacts of habitat degradation and increased parasitism. Vulnerable species are able to withstand parasitism where habitat tracts are extensive, because Cowbird parasitism drops off toward the interior of densely wooded areas. The reduction in the extent of riparian cover coupled with the increase in edge effect (where parasitism is higher), and the increase in the Cowbird population has crossed an environmental threshold that endangers these host species.

woodlands; open fields with rank annual grasses, fiddlenecks, and thistles; and a reliable water source. Northern Flicker is rarely found in the deep forest, but prefers the forest edge, where dead and dying trees and bare ground provide appropriate habitat. Bushtits in the Coast Ranges typically like hard chaparral intermixed with evergreens, particularly live oak and ceanothus, and where hanging lichen provides nesting material for their pendulous nests. The following description of Lazuli Bunting *(Passerina amoena)* habitat gives an idea of how necessarily complex an edge must be to attract this wide-ranging bird, which is always a pleasure to encounter:

> Clumps of bushes, broken chaparral, weed thickets and other low growing vegetation on hillsides or in and about water courses, but not usually over water or damp ground. . . . Diversity of plant growth and discontinuity of masses of it seem important as well as presence of low dense tangle used normally for nesting. (Grinnell and Miller 1944)

Dimorphism and Monomorphism

Anna's Hummingbird is sexually *dimorphic;* the plumage of the adults of each sex is distinct. The male is festooned with a bright magenta throat *(gorget)* and head; the female has a much less conspicuous plumage. This common hummingbird occurs across much of the Pacific slope, moving easily (and often) among habitats, from chaparral and coastal scrub through mixed woodland and riparian edges, exploiting a variety of nectar sources among seasons. Within a given species, males and females may have different habitat preferences, especially those species that are sexually dimorphic. For example, male Anna's Hummingbirds "tend to take territorial stands in the more open situations, as up canyon sides or hillslopes or out of level washes, while the females in their nesting activities adhere to tracts of evergreen trees, most commonly, perhaps . . . the live oak" (Grinnell and Miller 1944). This separation of habitat use is made possible by the fact that male Anna's Hummingbirds are polygynous; they mate with many females and take no part in nesting or chick rearing. By comparison, male and female Song

Plate 11. The brilliant crown and gorget of a male Anna's Hummingbird serves to advertise his presence to rivals and prospective mates in early spring. The widespread introduction of exotic fall-flowering trees and shrubs helps the Anna's Hummingbird to overwinter in California.

Sparrows are indistinguishable from each other; both wear the same plumage and are the same size. Likewise, each sex participates fully in nesting and chick rearing, and each occupies the same habitat niche. This egalitarian lifestyle holds true for most species: when both sexes look alike, when they are *monomorphic*, each tends to behave similarly and participate equally in their nesting efforts. When each sex is distinct in its plumage or morphology, behavior differs. Also, the greater the physical differences, the greater the behavioral differences. Sexual dimorphism is perhaps most pronounced in the Greater Sage-Grouse, the emblematic species of the Great Basin. (A discussion of Sage-Grouse "lekking" behavior is provided in "Birds of the Great Basin.")

Plate 12. The female Anna's Hummingbird is green-backed and gray-bellied with a speckled throat. All female hummingbirds are notoriously difficult to distinguish in the field.

Taxonomy: Superspecies, Species, and Subspecies

Taxonomy is the science of natural relationships among and within types of living things. The biological *species*—a genetically discrete group of populations—is the natural unit of taxonomy. But boundaries between species, like boundaries between habitats, are not always distinct. The species concept is ever evolving, especially with advances in DNA analysis, and distinctions that were formerly obscure are now possible. In ornithology, as in other natural sciences, the pendulum swings between periods of lumping and splitting. Lumping is the merging of two, or even several, species or subspecies into one. Splitting is the division of what was thought to be a single species into usually two, or possibly more, species. Because of the ongoing refinement of understanding of relationships within and among species, and because of the nearly constant modifications of the *Check-list of North American Birds* (American Ornithologists' Union 1957, 1983, 1998, and supplements), the number of species recognized by the Committee on Classification and Nomenclature of the American Ornithologists' Union is under constant revision.

The subspecies concept is equivalent to *geographical race.* Subspecies are "species-in-the-making" (American Ornithologists' Union 1998) because they represent geographic portions of species' populations that have discrete differences in morphology, coloration, and biochemistry, and often corresponding differences in habitat and behavior. The study of subspecies involves a complex and dynamic investigation within taxonomy; subspecies are living examples of evolution's subtlety and ingenuity, and they express the sensitivity of species to the environments in which they live.

Early California ornithologists, centered mostly around Joseph Grinnell and his associates at the University of California at Berkeley, were "splitters," paying close attention to the differences between races, especially those that showed wide variation in morphology among geographic locations. For example, Grinnell and Miller (1944) recognized 15 subspecies of Fox Sparrow *(Passerella iliaca)* as occurring in California and studied the plumage and distribution of each diligently. The importance of

Plate 13. Young Fox Sparrows, in a nest concealed under the tangle of a mountain thicket, respond to the arrival of a parent with food. Fox Sparrows are ground feeders and have the habit of scratching like chickens to find food.

subspecies was reflected in the American Ornithologists' Union's 1957 Check-list of North American Birds, which afforded each subspecies a thorough account. Subsequent editions (1983, 1998) of the Check-list were more species oriented, providing limited information on subspecies, but at the same time expanding the *superspecies* concept, which addresses relationships between closely related "sibling species," species that have diverged fairly recently but apparently maintain genetic integrity through geographic and behavioral isolation—for example, California Quail and Gambel's Quail *(Callipepla gambelii)*. The period of lumping that occupied the latter half of the twentieth century has recently given way to renewed interest in variation within species. Over the last decade or two, California's avifauna have gained several species through a splitting of the Western Grebe *(Aechmophorus occidentalis)* to recognize Clark's Grebe *(A. clarkii);* of "Western Flycatcher" into Pacific-slope and Cordilleran Flycatchers *(Empidonax difficilis* and *E. occidentalis);* of "Plain Titmouse" into Oak Titmouse and Juniper Titmouse; and of "Brown Towhee" into California Towhee *(Pipilo crissalis)* and Canyon Towhee (*P. fuscus;* the latter is not a California species, however). So, too, "sister species" such as California Gnatcatcher and Black-tailed Gnatcatcher *(Polioptila melanura)* have been recognized.

Incipient species, populations that are intermediate between races (subspecies) and species, are sometimes referred to as "semispecies." Closely related species that are considered too distinct morphologically to be a single species are called superspecies. Examples of California species pairs that constitute superspecies are Ladder-backed and Nuttall's Woodpeckers *(Picoides scalaris* and *P. nuttallii);* Rufous and Allen's Hummingbirds *(Selasphorus rufus* and *S. sasin);* Juniper and Oak Titmice; and Pacific-slope and Cordilleran Flycatchers. Species such as these that are each the other's closest relatives arose from a common ancestral species. Each member of each pair differs from the other enough in voice, morphology, coloration, and ecology to qualify as a sister species, and where their ranges overlap, where they are *sympatric,* they do not mate randomly with each other and are therefore reproductively isolated. The status of closely related species is not always so easy to sort out, however.

The Baltimore Oriole *(Icterus galbula),* an eastern species, and Bullock's Oriole *(I. bullockii),* a western species, were formerly lumped as a single species, the "Northern" Oriole. It had long been known that the two differed in male, female, and immature plumages and vocalizations, but subsequent study found distinct differences in gene frequencies, physiological response *(thermoregulation),* molting sequence and schedule, nest-site placement, and body size—all indicators that the gene flow between the two "subspecies" is limited. Therefore, they are now considered distinct species.

The two types of Northern Flicker—"Yellow-shafted" (eastern) and "Red-shafted" (western)—illustrate another pattern. These were formerly treated as separate species, but it was found that in their zone of contact (the Great Plains), interbreeding between the two groups was uninhibited. This random assortment created offspring that could not be identified as either "Yellow-shafted" or "Red-shafted," so the two were merged into the Northern Flicker. An analogous situation exists between "Audubon's" and "Myrtle" Warblers, which interbreed freely in the Canadian Rockies and thus have been lumped into the Yellow-rumped Warbler *(Dendroica coronata),* a common wintering species throughout the California lowlands.

Plate 14. Male and female Bullock's Orioles share in the feeding of their nestlings. The nest, usually hung from the outer branch of a tree, consists of a deep pouch woven with plant fiber and sometimes string, horse hair, or fishing line. The male shown here is in subadult plumage, typical of males in their first breeding season.

Hybridization among closely related species confounds the definition of species (see the glossary) but also sheds light on the nature of evolution among birds. When closely related, or recently diverged, species of the same *genus (congeners)* come into contact, hybridization may occur. Most of the species that do interbreed belong to a superspecies complex, and although they may interbreed to a limited extent in overlapping portions of their ranges, they are behaviorally isolated. (Pacific-slope and Cordilleran Flycatchers are sympatric in the northern Siskiyou Mountains, but they are isolated by behavioral differences.) Species pairs of California birds that are prone to limited hybridization include: Red-breasted Sapsucker *(Sphyrapicus ruber)* × Red-naped Sapsucker *(S. nuchalis);* Townsend's Warbler *(Dendroica townsendi)* × Hermit Warbler *(D. occidentalis),* Clark's × Western Grebes, Nuttall's × Ladder-backed Woodpecker, and possibly Nashville Warbler *(Vermivora ruficapilla)* × Virginia's Warbler *(V. virginiae),* among others. In contrast, Western Gull *(Larus occidentalis)* × Glaucous-winged Gull *(L. glaucescens)* hybrids are rather common, though mostly overlooked. Occasionally, hybrids are found that are the product of western (Californian) and mostly eastern species:

Lazuli Bunting × Indigo Bunting *(Passerina cyanea),* Black-headed Grosbeak *(Pheucticus melanocephalus)* × Rose-breasted Grosbeak *(P. ludovicianus),* and sparrows of the genus *Zonotrichia* (especially Golden-crowned *[Z. atricapilla]* × White-throated *[Z. albicollis]*).

Population Status

The avifauna can be divided into various categories based on abundance (rare to abundant), biogeographical history (extinct, extirpated, or *extant*), population trend (increasing, stable, decreasing), listed status (rare, threatened, or endangered), taxonomy (evolutionary relationships within and among species), and seasonal occurrence (year-round, winter resident, summer resident, transient). This book focuses on the largest group, the native common species, but an understanding of the other categories is necessary to understand California's birdlife.

Most avifaunal accounts use the standard modifiers—common, fairly common, uncommon, rare, and extremely rare—as general indicators of abundance. Because of natural differences in abundance according to a species' place in the food chain, some species will never be as abundant as others. For example, Dunlin *(Calidris alpina),* an Arctic-breeding shorebird, visits California's tidal flats in fall and winter. Several thousand Dunlin (a common species of coastal estuaries) may forage together across the flats and roost in tight flocks in the higher marsh or on sandbars. A bird of prey known as a Merlin *(Falco columbarius)* visits the same tidal flats, often perching on a post or pole, studying the shorebird flocks at some distance, calculating an attack strategy. The Merlin is usually classified as rare or uncommon, and it will always be so. The concept of the food pyramid tells us that predators must always occur in much smaller numbers than their prey.

Species that are undergoing population declines may be included on official lists produced and maintained by federal or state agencies or nongovernmental organizations. The highest degree of protection is provided for those species that are federally listed as threatened or endangered under the Endangered Species Act. The California Department of Fish and Game also lists species that are threatened or endangered within the state, although they may not be included on a federal list. An example is the California Black Rail *(Laterallus jamaicensis coturniculus)*, a subspecies of Black Rail restricted almost entirely to California that is listed as "threatened" by the California agency but which has no official listing at the federal level because it has not been demonstrated that the populations outside of California are at risk (although they may well be). Both the federal U.S. Fish and Wildlife Service and the state agencies also maintain a list of Special Animals (U.S. Fish and Wildlife Service) or Bird Species of Concern (California Department of Fish and Game) that identify populations suffering some decline, usually because of habitat loss, disturbance of nesting sites, or environmental contamination. These listings afford less protection than the "threatened" or "endangered" categories, but act as a red flag to regulatory and permitting agencies when projects are proposed that may impact the health and well-being of these species.

Additionally, the National Audubon Society, a nongovernmental organization devoted to the protection of birds and their habitats, periodically publishes a "Watch List" of species not included on the official lists, but considered to be at risk. Watch-listing is often a precursor, an early warning sign, to inclusion on a state or federal list.

Endemics and Near Endemics

Endemism, the condition of being restricted to a certain place, occurs at both the subspecies and species level. California hosts three fully endemic species—California Condor *(Gymnogyps californianus)*, Island Scrub-Jay *(Aphelocoma insularis)*, and Yellow-billed Magpie *(Pica nuttalli)*. The Condor is essentially extinct in the wild, though it has been recently reintroduced; its status is discussed below. The Island Scrub-Jay is restricted entirely to Santa Cruz Island (96 square miles) and differs from its closest relative on the mainland, the Western Scrub-Jay, by its larger size, brighter blue color, hoarser voice, and different breeding behavior. The Magpie, in contrast, ranges throughout the Central Valley (25,000 square miles) and along the south coast, but nowhere else except as a rare vagrant. Like the Condor in pre-European

Plate 15. The Yellow-billed Magpie is a California endemic. It is a common resident, found in the Central Valley and adjacent foothills and in the Coast Ranges from San Francisco Bay to Ventura County.

times, several other species are nearly endemic with the vast preponderance of their breeding populations occurring within the state—Ashy Storm-Petrel *(Oceanodroma homochroa),* Nuttall's Woodpecker, Allen's Hummingbird, Oak Titmouse, Wrentit *(Chamaea fasciata),* California Thrasher *(Toxostoma redivivum),* Tricolored Blackbird, and Lawrence's Goldfinch. Although the Storm-Petrel and the Hummingbird disperse widely after the breeding season, because their breeding range is within California and they likely evolved here, they are essentially endemic. Tricolored Blackbird barely spills over the borders of California, having small breeding colonies in southern Oregon and Baja California. Nuttall's Woodpecker is essentially a bird of California's live oak woodlands, but it also breeds in Baja California and wanders very rarely into Oregon and Nevada.

Geographic isolation is a double-edged sword; it can drive speciation, but it also can be an evolutionary cul-de-sac. The dangers of specialization are illustrated by the Ashy Storm-Petrel. The world population is essentially confined to only a few locations, the Farallon Islands (75 percent of the population) and the Channel Islands (the rest), where there are very limited breeding opportunities in rocky crevices. Most of these Storm-Petrels flock together in Monterey Bay in fall, potentially exposing the entire population to a single catastrophic event—environmental contamination, disease, or severe weather. The concentrated breeding colonies also face threats such as nest disturbance by humans and livestock; predation by introduced mammalian predators such as rats and cats; depredation by gulls, exacerbated by artificial lights near the colony (or on boats near the colony); and reduced breeding success or mortality because of oil spills. These threats hang over many of California's seabirds, but the Storm-Petrel, being so local and so geographically isolated, is especially susceptible.

Although only a few full species are truly endemic to California, subspecific endemism is rampant in California, which has 60 endemic subspecies; another 56 subspecies are near endemics. Remembering that subspecies are "species in the making," it is apparent that California is an evolutionary dynamo. A few examples illustrate the nature of subspecies and endemism. Perhaps the best example of a discrete population is the Inyo California Towhee *(Pipilo crissalis eremophilus)*, federally listed as threatened because fewer than 200 individuals are known to exist. The Inyo California Towhee is a relict population of a species that was historically widespread in the southwestern United States and northern Mexico. As a result of prehistoric climatic changes beginning in the Pliocene, this population became restricted to mountain areas in the northern Mojave Desert. Its range is now limited to riparian habitats within the southern Argus Range in Inyo County, where it is geographically isolated from other subspecies of the California Towhee. The population is dependent on a limited riparian habitat that has been reduced or eliminated by various human activities—grazing, export of water, mining, recreational and military activities, and rural development.

San Clemente Island, 49 miles from the mainland and 56 square miles in size, is large enough and isolated enough to spin off several subspecies, and until recently it hosted endemic subspecies of Loggerhead Shrike *(Lanius ludovicianus)*, Bewick's Wren *(Thryomanes bewickii)*, Spotted Towhee *(Pipilo macula-*

tus), Song Sparrow, and Sage Sparrow. Habitat disturbance and predation has left the island with only Sage Sparrow and the San Clemente Loggerhead Shrike *(L. l. mearnsi),* the latter a subspecies with perhaps the smallest range of any bird in North America as well as one of the most endangered. Although shrikes formerly occurred throughout the island, defoliation and depredation by feral cats and ravens have taken their toll.

Another endemic subspecies is the San Francisco Common Yellowthroat *(Geothlypis trichas sinuosa),* also know as the "Saltmarsh Common Yellowthroat." The Common Yellowthroat is rather widely distributed throughout marshes of North and Central America and Mexico, and more than a dozen subspecies are recognized; but the San Francisco Common Yellowthroat is mostly nonmigratory, with the breeding population confined to the San Francisco Bay Region. Differences between this subspecies and others are subtle but measurable and illustrate the kinds of morphological characters that distinguish subspecies—smaller size, shorter wing length, darker mantle, more brown on the flanks, and a more rounded wing. Behavioral differences may distinguish subspecies as well; for example, some Yellowthroat populations are nonmigratory, whereas others are migratory.

Species that include both migratory and nonmigratory populations tend to develop subspecies within their ranks. Nuttall's White-crowned Sparrow *(Zonotrichia leucophrys nuttalli)* resides within a narrow strip of coastal scrub habitat from Mendocino County southward to Point Conception, the "fog belt" of the

NUTTALL'S ENDEMICS

Two of California's endemic birds, Nuttall's Woodpecker *(Picoides nuttallii)* and Yellow-billed Magpie *(Pica nuttalli),* bear the specific Latin name *nuttalli* after the intrepid British botanist, Thomas Nuttall, author of the first North American field guide, *A Manual of the Ornithology of the United States and of Canada (1832–34).* Nuttall was the first European naturalist to visit California and was the first to collect specimens of not only the state's unique woodpecker and magpie but also the Tricolored Blackbird, in the spring of 1836 near Monterey. His preeminence in describing California's avifauna lives on in other names as well, for example, Common Poorwill *(Phalaenoptilus nuttallii)* was formerly known as "Nuttall's Whip poor will."

central coast. Formerly considered a full species, it is now recognized as a subspecies. In winter it is sympatric with the migratory Puget Sound White-crowned Sparrow *(Z. l. pugetensis)*, which retreats to breed from north of Mendocino to British Columbia. What separates these two populations? Although there are no dramatic geographical barriers, a physiological isolating mechanism results from differing breeding distributions. The seasonal development of the gonads, which is controlled by day length, differs in timing between the two populations; therefore, the more northerly Puget Sound population is not ready to breed when Nuttall's Sparrows begin their breeding season. Hence, a reproductive barrier exists between these two populations.

"Accidentals"

Of those more than 600 species of birds known to have occurred naturally in the Golden State, a large proportion, more than 25 percent, are exceedingly rare here, occurring only briefly as vagrants or transients. Records of many of these rarities are scrutinized by the "California Bird Records Committee," and they must be thoroughly documented by reliable observers to be accepted on the official state list. As might be expected, some groups of birds are much more prone to wandering beyond the boundaries of their normal distributions than others, a phenomenon known as vagrancy. (The term vagrant has none of the pejorative connotations applied to human vagrants. Indeed, an avian vagrant is a cherished find for active birders.) Certain avian families have a tendency to range far and wide and are more apt to occur as vagrants. Waterbirds, as a general group, tend to be best represented among the wanderers, both in terms of numbers of individuals and numbers of species. Of the tubenoses, those legendary oceanic wanderers that include the albatrosses, shearwaters, and storm-petrels, fully 42 percent (13 of 31 species) of those that have occurred in California are rarities. Although a large proportion of the tubenoses breed in the Southern Hemisphere, their ocean wanderlust brings them along our shores regularly. Representative species demonstrate the propensity of that family to wander from distant breeding grounds into California waters: Short-tailed Albatross *(Phoebastria albatrus)* breeds on Torishima and Minamiko-jima, which are islands off Japan; Cook's Petrel *(Pterodroma cookii)* nests on islands off New Zealand; and Wedge-tailed Shearwater *(Puffinus pacificus)*, which is exceedingly rare in California, nests in Hawaii and off Mexico.

Among the landbirds, the *nine-primaried ocines*, especially the wood-warblers, are most famously prone to vagrancy, and searching for such vagrants has become a popular pursuit for many California birders, neophytes and veterans alike. On any given weekend from late August through early October, on coastal promontories and known "vagrant traps," you are likely to find denim- and khaki-clad birders scrutinizing a windblown cypress tree or a clump of lupines at favored hot spots such as Trinidad Head, the Point Reyes Lighthouse, Pigeon Point, Carmel River mouth, Goleta, Point Mugu, Atascadero willows, or Point Loma. An unfamiliar call note, or a shadow of movement in the dark recesses of the cypress boughs, keys the observer's attention, and binoculars are raised to capture a glimpse of some diagnostic clue—an eye ring, a wing bar, or even leg color—combinations of characters that may help distinguish a Blackpoll from a Bay-breasted Warbler *(Dendroica striata* and *D. castanea)*. Or, perhaps, behavioral cues will be useful. Does the bird habitually flick its wings? If so, it may be a Hermit Thrush rather than a Swainson's Thrush *(Catharus ustulatus)*. Is the small warblerlike bird flicking its tail sideways? Yes? Probably a Prairie Warbler *(D. discolor)*, but best to get a look at its face pattern.

The role of vagrancy in the survival of species is a topic of particular interest to birders. In an earlier era in California ornithology, the term "accidental" was used to describe the status of extremely rare birds—those that had been detected only once or twice in the state. As records accumulated, the species might be reclassified as "casual," and then simply as "rare." Joseph Grinnell recognized early that the use of the term "accidental" was misleading, because there is nothing accidental, or haphazard, about the phenomenon of vagrancy, rather, it is part of the ordinary evolutionary program.

> It is obvious that the interests of the individual are sacrificed in the interests of the species. The species will not succeed in maintaining itself except by virtue of the continual activity of pioneers, the function of which is to seek out new places for establishment. Only by the service of the scouts is the army as a whole able to

> advance or to prevent itself being engulfed: in the vernacular, crowded off the map—its career ended. (Grinnell 1922)

But the intrepid birder should keep in mind that vagrancy, by its very nature, is a rare event, so vagrants are rarely encountered. The remainder of this book focuses on California's "usual," or emblematic avifauna; but it is instructive to understand the evolutionary dynamism that is expressed in the biogeography of birds, and the reality that nothing is certain but change.

Introduced Species

Of the total list of California birds, the following nine are introduced, nonnative species:

Chukar, *Alectoris chukar*
Ring-necked Pheasant, *Phasianus colchicus*
White-tailed Ptarmigan, *Lagopus leucurus*
Wild Turkey, *Meleagris gallopavo*
Rock Pigeon or Rock Dove, *Columba livia*
Spotted Dove, *Streptopelia chinensis*
Red-crowned Parrot, *Amazona viridigenalis*
European Starling, *Sturnus vulgaris*
House Sparrow, *Passer domesticus*

An additional 15 nonnative species have established populations in California, but whether these are viable breeding populations is unresolved in most cases. Four nonnatives—Chukar, Pheasant, Ptarmigan, and Turkey—were introduced as game birds; all but the Ptarmigan have become fairly well established and widely distributed. Attempts to introduce nonnative, or alien, game birds into California were ill conceived, and the consequences have been mostly negative, resulting in displacement of native species, introduction of disease, and alteration of natural habitats and plant communities. However, these liabilities have done little to dampen the effort of introducing foreign species, and it has not been limited to birds. Dramatic examples of the unintended consequences of introduced animals are provided by fish, amphibians, and mammals. Red Fox *(Vulpes fulva)*, inadvertently introduced into coastal lowlands of California, has

Plate 16. By displaying his stunning wattles, a cock Wild Turkey maintains dominance over his hens. Turkeys have been "planted" as a game species in California, but there is some concern over their impact on the regeneration of oak forests.

had a devastating effect on California Clapper Rail, Western Snowy Plover *(Charadrius alexandrinus nivasus),* and other ground-nesting birds. The introduced European Starling has displaced many cavity-nesting birds from nesting sites, perhaps most seriously the Western Bluebird *(Sialia mexicana).* An introduced sports fish, Tilapia *(Oreochronis mossambicus),* is the vector of a botulism pathogen in the Salton Sea that has caused the death of hundreds of thousands of waterbirds since 1996 (see "The Desert's Birds").

Many efforts were made, especially early in the twentieth century, to introduce a variety of game birds—Hungarian ("Gray") Partridge *(Perdix perdix),* Northern Bobwhite *(Colinus virginianus),* Chinese Quail *(Coturnix chinensis),* Elegant Quail *(Callipepla douglasii)*—but they met with failure. Perhaps the most successful nonnative game bird established in California is Ring-necked Pheasant. By 1916, the Fish and Game Commission had released more than 5,000 pheasants in more than half the counties of the state (Grinnell et al. 1918); today, they are widespread in lowland grasslands.

Wild Turkeys were introduced to Santa Cruz Island as early as 1877; these birds reproduced for a few years, but gradually the population decreased in size and ultimately died off. In 1908, the Fish and Game Commission imported 22 Wild Turkeys and 11 Chachalacas *(Ortalis vetula)* from Mexico and released them at the elevation of 4,000 feet in the San Bernardino Mountains; more turkeys were released in the same area in 1910, but none were seen again. Subsequent releases at Yosemite and Sequoia National Parks were more successful, but they mixed with barnyard turkeys and became semidomesticated, a creature entirely different in habitat and behavior than the noble bird of the eastern forest that Ben Franklin nominated as the national bird. But Fish and Game persisted, and ultimately succeeded, in establishing wild breeding populations in California; the stock now present is semiferal, and seems to do best in association with oak woodland habitats of the Coast Ranges and Sierra foothills, where they thrive on acorns. The state's largest concentrations of Wild Turkeys are found in Butte, Calaveras, El Dorado, Mendocino, Nevada, San Luis Obispo, Shasta, Tehama, and Yuba Counties, and Turkeys are increasingly found along the coast as well, especially in Marin and Sonoma Counties. The effects of Wild Turkeys spreading into formerly unoccupied oak woodland, coastal scrub, and forest are not well understood. The consumption of acorns is apparently inhibiting oak regeneration in some areas, but impacts to native animals may also be an issue. California Quail, the state bird, may be displaced from turkey-occupied territories, and small animals such as the California Slender Salamander *(Batrachoseps attenuatus)* may suffer depredations by the voracious flocks. The spread of Turkeys beyond their natural boundaries is actively aided and abetted by the Department of Fish and Game, an agency entrusted with the protection of the native flora and fauna. Wildlife biologists, botanists, and concerned environmental groups are now calling into question the wisdom of such game-bird introductions and requesting that control measures reverse the trend of expanding Turkey populations into California's wildlands.

Other attempts have been made to improve on nature's judgment and transplant native species outside the confines of their natural range. Efforts to transplant the desert-adapted Gambel's Quail to more northerly portions of the state were spectacularly unsuccessful. As an example, in 1907 a "large number" of Gambel's Quail were released at a country club in coastal Marin County, but two years later none could be found. Fortunately, there is no evidence that crossbreeding with the California Quail resulted.

Expatriates

In deference to their tragic histories, we end our "Overview of California Birdlife" with a brief tribute to two native species that disappeared from the state precipitously following the arrival of Europeans. Within historical times, the California Condor *(Gymnogyps californianus)* and the Sharp-tailed Grouse *(Tympanuchus phasianellus)* have been extirpated or lost to the wild. The Grouse still occurs, though in diminishing numbers, in some portions of its range outside of California, from central Alaska southward through western Canada to eastern Washington and Oregon and elsewhere, in prairie grasslands, scrub, and *muskeg.* The conversion of its California range from shrublands and perennial bunchgrass to largely annual grasslands has, sadly, rendered this state inhospitable to the Sharp-tailed Grouse.

Early naturalists such as Cooper, Henshaw, and Bendire, who described the fauna of California shortly after the Gold Rush, found the Sharp-tailed Grouse quite common in the Modoc region northeast of the Cascade Range. The "prairie-chicken" was once plentiful enough "to afford excellent shooting" (Henshaw 1880), and the legislature proclaimed open season on prairie-chickens in Siskiyou County in 1866. It is no surprise, then, that by the early 1900s their numbers had diminished greatly because of relentless hunting and conversion of their favored grasslands to heavily grazed ranchlands. Perhaps this early diminution in California's avifauna was not for naught:

> It is hoped that in the history of the Columbian Sharp-tailed Grouse, California has learned a lesson that will result in benefit to every other wild species in the state. Here is a magnificent game bird, completely eliminated from our confines as a result of unrestrained hunting. A modicum of foresight and forebearance would doubtless have preserved the bird. (Grinnell et al. 1918)

A valiant, expensive, and perhaps quixotic effort is under way to reintroduce the Condor to parts of its former range; but the conversion of California from a prairielike province populated with antelope and perennial grasslands to an agricultural stockyard, and the disappearance of large herds of hoofed mammals on which the Condor once fed, has relegated this noble master of the sky to the sad status of an anachronism. That said, well-cared-for and tenderly reared Condors may be seen in the wilds of Big Sur or the Los Padres National Forest, all sporting bright plastic wing tags.

The Condor once scribed the California sky west of the Sierran crest in search of dead game—a stillborn antelope, a fallen mountain goat, deer, or coyote. The earliest explorers describe these sky masters glutting on whale carcasses on the shore at Monterey or appearing from the clouds to descend on a recently killed grizzly. According to early frontiersman John Clayman, you could hear the wind singing through their flight feathers a mile away as they swooped down on a fresh carcass. Now the Condor is raised in captivity, attended and coddled by a host of dedicated biologists, to be released into "the wild," in an effort to resurrect the fallen giant. The truth be known, California, like the rest of North America, has become too civilized, too small really, to accommodate the freedom embodied by its most magnificent residents—the Grizzly Bear *(Ursus arctos horribilis)* and the California Condor.

> When the ancient Gymnogyps, greatest of all flying birds of the earth, shall have passed forever from the ken of man, leaving in the lonesome vistas of its homeland mountains a dreary void no living creature can ever fill, the moment will be opportune for a more competent digest of these annals in their entirety. That time is drawing nearer year by year, and cannot be far off. (Harris 1941)

Fortunately, despite the sad losses recounted above, in almost every location and during every season, California is endowed with an abundant avifauna, providing ample opportunity for enjoyment of its richness and beauty—whether searching the frosted Sierran meadows for a Great Gray Owl, the autumn estuary for a Black-bellied Plover in remnant breeding plumage, or the spring willow thicket for the effervescent song of a Yellow Warbler.

SEABIRDS AND THE MARINE ENVIRONMENT

The Pacific—widest, deepest, longest, and most productive of the great oceans. . . . At the mercy of wind and wave, the marine birds of this and all seas are held in a tenuous truce with the earth's most powerful forces. As the sea goes, so must they, and their existence depends upon its mood, health, and productivity.

DELPHINE HALEY, *SEABIRDS OF THE EASTERN NORTH PACIFIC AND ARCTIC WATERS*

Under compelling influences of ocean currents emanating from both the Gulf of Alaska and the subtropical Pacific, and with a generous scattering of offshore islands and a complex submarine topography, the nearshore and oceanic waters of California offer myriad opportunities and daunting challenges for seabirds. The varying effects of wind, current, and water temperature interact with season to determine availability of prey, and, in turn, the abundance and distribution of California's seabirds, especially for those species that rely on waters farther offshore. Following a spring when the wind has been strong and prevailing out of the

Map 2. Marine Environment Bioregion.

northwest, the upwelling cycle has persisted, and the water has been cold, *primary productivity* of the ocean is high, and both breeding and visiting species find abundant food. In contrast, during El Niño periods, when northerly winds weaken, subtropical currents predominate, and water temperatures are relatively warm, prey becomes scarce and seabirds suffer, either through die-offs or lower reproductive success.

Through long-term monitoring of seabird colonies along the coast, especially at the South Farallon Islands and San Miguel Island, frequent pelagic birding and whale-watching excursions, and data-gathering oceanographic cruises, biologists are gaining a fuller understanding of the cycles of abundance and scarcity that characterize the California Current. This chapter provides an overview of the seabird communities that rely on California's waters for sustenance and considers the adaptations seabirds have made to the unique oceanic conditions of the nearshore eastern Pacific.

California's Marine Environment

Bathymetry and Topography

The coastline of California, extending 10 degrees of latitude (32°N–42°N; 1,100 miles) between its borders with Oregon and Mexico, has two distinct oceanographic regions defined largely by its geography. The northern two-thirds of the coastline trends in a southerly direction from Oregon, then breaks southeasterly at Point Conception, a swollen knuckle just north of Santa Barbara. Point Conception is the dividing line between two regions, each of which is characterized by unique oceanographic conditions that support a somewhat different avifauna. Northward of Point Conception, the cool, upwelling influence of the California Current dominates; southward, the waters are within the Southern California Bight, where warmer currents originate in subtropical latitudes.

From any of the promontories along the length of the state—Point Saint George, Trinidad Head, Cape Mendocino, Point Arena, Point Reyes, Point Sur, Palos Verdes, or Point Mugu—the surface of the ocean may seem like a uniform expanse of open

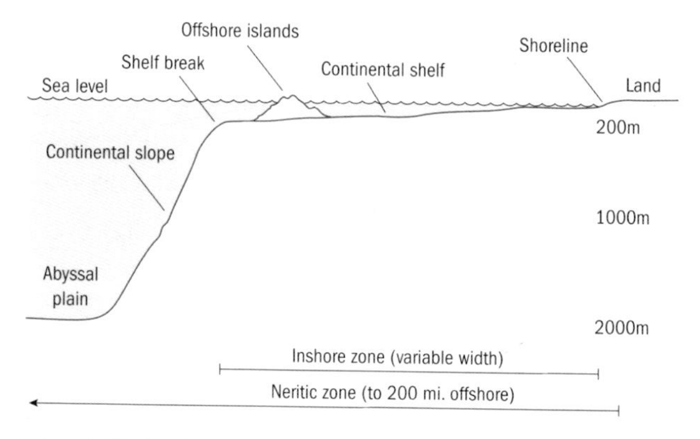

Figure 1. Idealized cross section of the neritic zone of the Pacific Coast showing major geomorphic features.

water corrugated with an endless series of even swells. But beneath the surface, the ocean floor varies greatly, and this submarine topography helps determine the patterns of current and water temperature that ultimately determine the distribution and abundance of marine birds.

Most of our knowledge of oceanic birds is confined to what is commonly called the *nearshore,* or *neritic,* zone, which refers to waters within 200 miles of the shore. Within that area, we refer to waters from the shore to the continental shelf break as "inshore," and those beyond the break, where the ocean floor drops off precipitously to the abyssal plain, as "offshore." The continental shelf is generally a region of gently sloping ocean floor between the shoreline and the continental shelf break. The shelf's width varies from as few as three miles, for example, at Monterey Bay, to as many as 45 miles, as off the Golden Gate and Eureka. Overall, the shelf is relatively narrow as compared to its width on the Atlantic coast of North America. Like California's landscape, the submarine topography of the continental shelf is relatively heterogeneous, a function of California's complex geologic history and tectonic dynamism.

The continental shelf break, that boundary between the continental shelf and the slope, is not a straight line but is as erratic and irregular as any mountain range. The continental shelf off southern California has been transforming itself into a basin and range topography for 30 million years, and, as a result, a variety of submarine canyons, escarpments, sea mounts, and submerged ridges interrupt, or intrude into, the continental shelf. Islands, important nesting habitat for seabirds, occur where submarine ridges break the surface. The Channel Islands group off Santa Barbara is an expression of the chaotic geologic history of rising and falling sea level, colliding continental plates, and roller-coaster tectonics that have formed the rifted borderland of the Southern California Bight. North of Point Conception, the continental shelf is more homogeneous, but the steep submarine topography generates colder, nutrient-rich upwelling waters that provide some of the world's most productive seabird habitat. The Farallon Islands, about 20 miles west of San Francisco, and the geologically related Cordell Bank are perched on the shelf break; the Farallon Islands support one of the densest colonies of breeding seabirds in the world, more than one-half million birds on its 100-acre main island, and the Bank attracts a diverse array of marine birds and mammals at all seasons. Monterey and Ascension submarine canyons cut close to shore at Monterey Bay, and the abrupt change in bottom topography, with depths of 2,000 feet within a few miles of the fishing docks of Monterey Bay, also provide a birding hot spot.

Oceanography: The California Current and the Southern California Bight

> *The ocean waters of the Pacific adjacent to the coast of California provide some of the finest and most easily accessible bird and marine mammal watching in the world.*
>
> RICH STALLCUP, *OCEAN BIRDS OF THE NEARSHORE PACIFIC*

The California Current is the dominant oceanic regime of the eastern Pacific Ocean, and it exerts a crucial influence on the seabirds that rely on its productivity for survival. The California Current is one of five of the world's "eastern boundary currents," each of which constitutes the eastern edge of a large oceanic *gyre* and flows along the western edge of a continent. Although eastern boundary currents account for less than 1 percent of the world's oceanic waters, they support one-third to one-half of the world's fisheries and a commensurate proportion of populations of other marine organisms.

Recent research has clarified the dynamics of upwelling systems, and the direct impacts of vacillating ocean temperatures on ocean productivity and marine animals are now fairly well understood. To understand seasonal occurrence and abundance of California's marine birds, it is essential to first understand the oceanography of the coast and the dynamics of the California Current and how it interacts with California's coastline. But the California Current it not easily fathomed.

The California Current is born in the Gulf of Alaska as an offspring of the North Pacific Gyre, a vast system of oceanic currents generated by planetary atmospheric winds, including the northwesterly winds that are so prevalent along the coast of northern California. The winds drive the surface waters of the eastern Pacific in a southeasterly direction, diagonally across the North Pacific Ocean and parallel to California's shoreline. This 100-mile-wide swath of surface water is, in turn, deflected offshore as it travels south because of the *Coriolis effect*. These surface waters, somewhat warmer than subsurface waters, are replaced by colder water from greater depths, a process known as *upwelling*. The cold, nutrient-rich waters associated with the upwelling phenomenon are more saturated with oxygen than are warmer waters, and they are therefore highly productive, supporting a rich community of plankton, the basis of the marine food chain.

Upwelling is not uniformly distributed, however. From the Farallon Islands northward to Eureka, water temperatures vary widely between cold upwelling fingers and warmer back eddies. The thermal fronts created by these variations often concentrate bioproductivity—blooms of phytoplankton and zooplankton that support the marine food web. In general, the strongest upwelling occurs along the middle portion of the state, from Point Conception to Cape Mendocino. This stretch of coastal topography is also intersected by several submarine canyons—Mendocino, Monterey, and Ascension—that bring very deep water close to shore. California's largest seabird rookery, the Farallon Islands, is situated dead center in the current's path and close to submarine topography—the continental shelf break and the Cordell Banks—which, in turn, generates its own upwelling currents.

As diverse as the avifauna of the California Current is in terms of number of species, it is inequitable in terms of abundance;

only three species—Sooty Shearwater *(Puffinus griseus)*, Common Murre *(Uria aalge)*, and Cassin's Auklet *(Ptychoramphus aleuticus)*—account for more than 90 percent of its seabird abundance and *biomass*. South of Point Conception, where the coast veers southeasterly, the influence of the northwesterly wind and the upwelling system diminishes, and warm subtropical atmospheric and oceanic currents ameliorate the effects of upwelling, at least in some seasons. The area under the spell of these more temperate, even subtropical, influences is known as the Southern California Bight. Warmer and less nutrient rich than the northern waters, southern waters support a somewhat different community of marine animals, including birds. But, of course, the Bight is not independent of the California Current. South and east of Point Conception is a "cyclonic recurvature" known as the Southern California Eddy, a cool water mass that transports upwelled water away from Point Conception and bounds the subtropical waters of the Southern California Bight close to the coast. The boundary between the cool waters of the Southern California Eddy and the warm waters of the Bight tend

Plate 17. As with many birds, pair bonding between Common Murres is strengthened by mutual preening.

to lie just east of San Nicolas Island. Upwelling occurs in the Bight as well; but its reliability and strength are reduced.

Recognizing that back eddies, countercurrents, and anomalies are integral to California's marine environment, an understanding of the generalized pattern of seasonal currents is helpful in explaining the occurrence of seabirds within these waters. There are three main seasons: the upwelling period, dominant from March into August; the oceanic period, from August through October, when upwelling subsides and warmer surface waters from offshore move coastward; and the Davidson Current period, from November through February, when subsurface waters move from south to north, between the California Current and the coast. The impact of these oceanographic seasons on California's biological productivity, and thus its seabirds, is critical. The abundance of phytoplankton declines during the oceanic period, reaches the lowest concentrations during the warm Davidson Current period, and then increases exponentially through the upwelling period (Ainley and Boekelheide 1990).

El Niño and the Southern Oscillation

An anomalous oceanographic phenomenon known as El Niño, the product of the southern oscillation, which is a periodic weakening of the atmospheric pressure gradient between the Pacific high and the Indian Ocean low pressure system, overrides the normal circulation patterns and water temperatures of the Southern California Bight and the California Current irregularly at two- to seven-year intervals. The dynamics of these shifting atmospheric cells is global in its amplitude, and its complexity is beyond the scope of this book. El Niño, however, has a profound effect on the distribution, abundance, and survival of seabirds in California. In short, the weakening pressure gradient and the equatorial trade winds create a long-period wave, called the Kelvin wave, that moves through the ocean at the depth where the temperature changes, called the *thermocline*. This deep wave sloshes northward along the coast of California and causes the overlying surface water layer to become deeper and warmer. This

Plate 18. Brandt's Cormorants are found exclusively along the coast or at sea, often in large feeding flocks. The limited availability of suitable nest sites on rocky promontories and offshore islands results in relentless competition and tight spacing between nests.

warm water affects the overlying air and, in turn, lessens the winds that drive upwelling.

This feedback loop between ocean and climate has become more thoroughly understood in recent years, and some of its impacts, particularly on breeding seabirds of the California Current, are well documented. These effects include a delayed onset in nest building and egg laying (cormorants, murres, Cassin's Auklet), hatching and fledging of fewer young, lower weights of those chicks that do fledge, higher adult mortality, emboldened predators (Western Gulls *[Larus occidentalis]* on Cassin's Auklets), and shifts in species' distributions and foraging strategies, among others. In years of extreme and sustained warm water, entire colonies may abandon their nests, as did Brandt's Cormorants *(Phalacrocorax penicillatus)* at South Farallon Island during a strong El Niño event in 1983. Or, an entire colony may fail to lay eggs, as did Pelagic Cormorants *(P. pelagicus)* in 1978 and 1983. Some species with more catholic foraging behaviors, and therefore more options available to them, fare better during El Niño years; for example, Western Gulls relied more on foraged garbage during 1983 and managed to reproduce in normal numbers.

So significant is the El Niño–Southern Oscillation phenomenon that it has been identified as a primary factor limiting the population size of seabirds, as well as other marine animals, through massive die-offs and reproductive failure. El Niño episodes probably strongly influenced seabird populations well before the industrial era, but in modern times, other factors exacerbate the natural cycle of die-off and reproductive failure. As seabird biologists David Ainley and Bob Boekelheide (1990) point out in *Seabirds of the Farallon Islands*, "overfishing has destroyed the entire food web, with the result that seabirds are now much fewer and apparently limited by food resources." Other factors also limit seabird populations. Availability of nesting sites has declined since prehistoric times because humans have occupied some islands where seabirds nested formerly. With human presence, populations of cats and rats increase, both of which are predators of ground-nesting seabirds. Human garbage serves to

Plate 19. Pelagic Cormorants, distinguished by their slender bills, favor nest sites on inaccessible rocky cliff ledges overlooking their oceanic habitat. Pelagic Cormorants are deep divers; there are accounts of their being caught in fishing nets at depths of 80 fathoms (480 feet).

Plate 20. Plastic debris, such as this discarded six-pack holder around the neck of a Western Gull, is a constant menace to seabirds.

subsidize gull populations, and gulls are aggressive and persistent predators of both seabirds and their eggs.

There is concern in the scientific community that El Niño events may increase in intensity and frequency as the greenhouse effect progresses and the climate changes. Warming of the California Current in recent decades has been identified as the cause of declines in populations, or distributional shifts, in a variety of marine organisms, especially fish and seabirds. El Niño phenomena may be related to the toxic algae blooms that have been found recently in Monterey Bay, where domoic acid (a potentially lethal neurotoxin produced by microscopic diatoms) was detected in high concentrations in anchovies, sardines, and krill—key prey species in the marine food web.

Taxonomic Groups

Seabirds come from the older, more "primitive" families of birds, those groups that have evolved over tens of millions of years in response to an environment that is relatively stable compared to that encountered by landbirds. The challenges imposed by the marine environment have produced some similarities among species that arose from families or orders related only distantly. In general, oceanic species tend to lack brightly colored plumage, tending toward subtle shades of gray, brown, and black. Many species are darker above and lighter below, an adaptation that may make them less visible to fish from below and aerial predators such as skuas and jaegers from above. Brighter colors are usually confined to the bill, eyes, legs, or *gular pouch,* and these increase in brightness during the breeding season. Most seabirds have very dense feathers and a well-developed oil gland *(uropygial gland),* which aid in waterproofing. Seabirds must also deal with the lack of freshwater, so they have developed special glands, located near the eyes, that are more efficient than kidneys for the desalination of blood. Some species, all "tubenoses" for example, excrete excess salt through external nostrils; others, such as pelicans and cormorants, have internal nostrils and so excrete salt crystals through their mouths. Although anatomical adaptations vary widely among seabird species (wing shape, for example), some features are common to most species. Seabirds tend to have the legs set far back on the body, where they function more like rudders or propellers than like limbs designed for walking. Feet are usually webbed, or sometimes lobed, as in grebes and phalaropes. Deep-diving species have denser bones than most birds; this adaptation lends ballast and reduces buoyancy. Divers also have higher blood volume and a higher capacity to retain oxygen for longer periods. The skeletal structure of many divers is also modified to withstand the pressure at depth.

Seabirds are among some of the most far-ranging animals in the world; it is not uncommon for a Sooty Shearwater or an Arctic Tern *(Sterna paradisaea)* to travel more than 20,000 miles between breeding seasons. Therefore, they have evolved exceptional navigation and orientation skills through mechanisms not fully understood, yet probably arising from innate abilities to draw upon a variety of clues including celestial orientation, the earth's magnetic fields, olfactory stimuli, low-frequency (wind-generated) sounds, and genetic memory.

Despite the common adaptations discussed above, all the seabirds that occur in California belong to only three taxonomic orders. An order is a rank within a hierarchy of classification and is the principal category between class and family. Thus, Sooty Shearwater is in the class Aves (birds), the order Procellariiformes (tubenose swimmers), and the family Procellariidae (shearwaters and petrels). Each order comprises several families, and its classification is based on genetic and morphological characters common to each family it includes. Within each order, groups are refined into more closely related genera and species. Other categories—superfamily, subfamily, and tribe—are refinements beyond the scope of this book, or most birders' concerns.

Procellariiformes: Albatross, Petrels, Shearwaters, and Storm-Petrels

The Black-footed Albatross *(Phoebastria nigripes),* by far the most common of the large tubenoses, patrols the coolest waters of the continental slope year-round, most commonly during the summer months and most commonly northward from Monterey Bay. Laysan and Short-tailed Albatross *(Phoebastria immutabilis* and *P. albatrus)* are both rather rare. Laysans tend to occur beyond the shelf break in open ocean, and before 1900, Short-taileds were rather common close to shore and even in San Francisco and Monterey Bays; but the decimation of nesting colonies in Japan has reduced their numbers worldwide. In-

Plate 21. The long, narrow wings of the Laysan Albatross enable it to wander widely across the Pacific Ocean.

creased sightings of Short-tailed Albatross in recent years, however, are encouraging.

Shearwaters occur in throngs, often of several species. Sooty Shearwater is the most abundant seabird in California waters, especially during cool-water years. When large flocks of Sooties congregate, they may be joined, most commonly, by Buller's, Pink-footed, and Flesh-footed Shearwaters *(Puffinus bulleri, P. creatopus,* and *P. carneipes).*

Black-vented Shearwaters *(P. opisthomelas)* nest on islands off Baja California and winter along nearshore escarpments off southern California in the tens of thousands, where they are often visible from shore; the Palos Verdes Peninsula is a good vantage point from October through April. Black-vents, uncommon north of Point Conception and casual north of Point Reyes, are associated with the Davidson Current and become more numerous and move farther north when it is more strongly developed.

Several other species of shearwaterlike tubenoses are rarities in offshore expanses of ocean and inspire a few intrepid naturalists into pursuit. These include, most notably, Mottled, Murphy's, Cook's, Dark-rumped petrels *(Pterodroma inexpectata, P. ultima, P. cookii,* and *P. phaeopygia/sandwichensis)* and Streaked Shearwater *(Calonectris leucomelas).*

The storm-petrels, the smallest of the tubenoses, glean small organisms—zooplankton and *micronekton*—from the ocean's surface with their delicate bills. The Leach's Storm-Petrel *(Oceanodroma leucorhoa)* is one of the most abundant seabird species in the world, whereas the Ashy Storm-Petrel *(O. homochroa),* nearly endemic to California, is one of the rarest; an estimated 85 percent of the world population nests on South Farallon Island. Leach's is broadly distributed and wide-ranging over the open ocean, whereas Ashy has a very limited distribution and stays close to the shelf break, rarely wandering farther offshore or closer inshore.

Pelecaniformes: Tropicbirds, Boobies, Pelicans, Cormorants, and Frigatebirds

Of the *totipalmate* swimmers (Pelecaniformes), only the pelicans and cormorants are true Californians; the other members of the group are primarily tropical and subtropical seabirds.

The Brown Pelican *(Pelecanus occidentalis)* is entirely marine,

Plate 22. Brown Pelicans are found along and outside the surf line and rarely venture inland. They feed by plunging from a height into schools of fish.

whereas American White Pelican *(P. erythrorhynchos)* is primarily a bird of inland lakes (see "Birds of the Great Basin"). Brown Pelican populations have a checkered environmental history, having suffered depredations by predators—both animal and human—and pesticides. In recent years, however, thanks to protection and a reduction in lethal contaminant levels, they are showing faltering but promising signs of recovery. Although Brown Pelicans nested formerly on other of the Channel Islands, and possibly even in San Francisco Bay, only two colonies remain, at West Anacapa and Santa Barbara Islands.

Although they do breed in coastal areas, Double-crested Cormorants *(Phalacrocorax auritus)* are birds more of estuaries and inland freshwater bodies and are discussed in "Birds of the Shoreline." The other two cormorant species—Brandt's and Pelagic—are truly marine species, however. Research on the Farallon Islands has found these two species to be highly sensitive to El Niño, and numbers found in the California Current have de-

Plate 23. The Red-billed Tropicbird, an occasional straggler into the seas off southern California, is pelagic and seldom seen near the coast.

creased since the 1970s. Both species have increased in the Southern California Bight over the same period; but threats from human disturbance, gill-net fisheries, and oil pollution continue.

Tropicbirds, boobies, and frigatebirds are essentially birds of tropical and subtropical latitudes and are relatively rare in California waters, although all groups are represented sporadically, especially toward the south. Red-billed Tropicbird *(Phaethon aethereus)* is the only one of the three tropicbird species that occurs regularly in California, and only offshore of the Channel Islands. Tropicbirds wander widely, however, especially during the postbreeding season, and each species has occurred in coastal California, albeit sparingly and far afield of its normal tropical distribution. Likewise, boobies, though common in Mexican waters, are rare in California, but five species (Red-footed, Blue-footed, Masked, and Brown Boobies [*Sula sula, S. nebouxii, S. dactylatra,* and *S. leucogaster*]) are possible, most commonly toward the south in fall and winter.

Charadriiformes: Shorebirds, Larids, and Alcids

Any similarity between a Red-necked Phalarope *(Phalaropus lobatus),* a South Polar Skua *(Catharacta maccormicki),* and a Common Murre may be lost on the casual observer, but not on the perspicacious taxonomists who have noted the commonality of tendon attachment and wishbone morphology among these disparate species. Perhaps no other avian order has diversified as greatly, or been as adept at carving out habitat niches, as the Charadriiformes.

Of the shorebirds, only the phalaropes have adapted to the open ocean. Red Phalarope *(Phalaropus fulicarius)* and Red-necked Phalarope are both Arctic breeders, and both species migrate southward after the brief Arctic breeding season. Both migrate en masse along California's continental slope, but Red Phalarope is the more oceanic of the two species and is rarely seen from land except when winter storms cause onshore "wrecks" and hundreds or even thousands of these plump pelagic shorebirds appear in coastal bays and beaches. On the occasion of one such November storm in 1976, Dave Shuford and I estimated 10,000 Red Phalaropes from one vantage point on Tomales Bay. Red Phalarope is one of those "instantaneous" species that swell the ranks of seabirds present in California during their passage en route to and from their breeding grounds in the Arctic and their wintering waters off the coast of South America. One-half million have been estimated in a single day passing by Pigeon Point. Red-necked Phalarope is more frequently seen from land, however, especially during fall migration.

Plate 24. Red Phalaropes are birds of the open ocean, but when inclement weather drives them inshore they can be seen feeding in coastal lagoons, pirouetting in the manner typical of phalaropes.

Plate 25. A Rhinoceros Auklet makes its way back to its nest following a nocturnal foraging trip. With protective management, these auklets are now recolonizing several offshore islands after a hiatus of many decades.

Among the larids, which include skuas, jaegers, gulls, terns, and skimmers, several species are truly pelagic, rarely wandering inshore of the shelf break. These oceanic wanderers include South Polar Skua, Pomarine and Long-tailed Jaegers *(Stercorarius pomarinus* and *S. longicaudus),* Black-legged Kittiwake *(Rissa tridactyla),* Sabine's Gull *(Xema sabini),* and Arctic Tern. In contrast to the alcids (below), these wide-ranging wanderers are also transequatorial, crossing the low latitudes with ease. Sabine's Gull and Long-tailed Jaeger are perhaps the most stubbornly oceanic of these species; none veer inside the shelf break with any regularity but steer seaward in their migration. Bonaparte's Gull *(Larus philadelphia),* too, can occur in vast flocks far offshore during migration, but also inshore and even inland. Most of the other species are distributed over inshore waters. Parasitic Jaeger *(Stercorarius parasiticus)* is most often encountered closer to shore, singly, as it marauds migrating flocks of Elegant Terns *(Sterna elegans)* along the immediate coast or even in larger estuaries.

The alcids, which include murres, murrelets, auklets, puffins,

Plate 26. Pigeon Guillemots produce weak but incessant calls, routinely resting on rocky promontories from which they can launch into flight.

and guillemots, are a product of the North Pacific, where the family has been evolving and refining its members since the early Eocene, for about 40 million years. The Northern Hemisphere's ecological and evolutionary equivalent of the Southern Hemisphere's penguins, alcids personify trade-offs in evolutionary forces between flying through the air and flying through the water. The alcids are submarine torpedoes, designed to forage by diving and pursuing fast-moving prey. The Common Murre, seen in flight, resembles a poorly thrown football; bottom heavy, labored, and struggling to stay aloft. Underwater, however, murres display the hydrobatic agility necessary to capture fast-moving schools of anchovies and smelt. This ability has allowed murres to become one of the most successful and abundant breeding species in the California Current. As noted above, of the 11 seabirds breeding on the Farallon Islands, five are alcids: murre, puffin, Cassin's Auklet, Rhinoceros Auklet *(Cerorhinca monocerata)*, and Pigeon Guillemot *(Cepphus columba)*. The Marbled Murrelet *(Brachyramphus marmoratus)*, also an alcid, has a unique habitat preference; although a bird of the open ocean, it nests in old-growth, coastal conifer forests. The wholesale logging of this habitat now threatens this murrelet's existence.

Biogeography of Seabirds

There are patterns to seabird distribution, and each seabird species has an affinity with some pattern of oceanic conditions. The marine birds can be divided into two main groups based on their biogeography: those that nest in California and those that visit only as immature birds or nonbreeding migrants. The 30 or so nesting species display a variety of geographical affinities indicative of California's equitable climate, productive waters, and the temperate latitudes it spans. There are three types of breeding

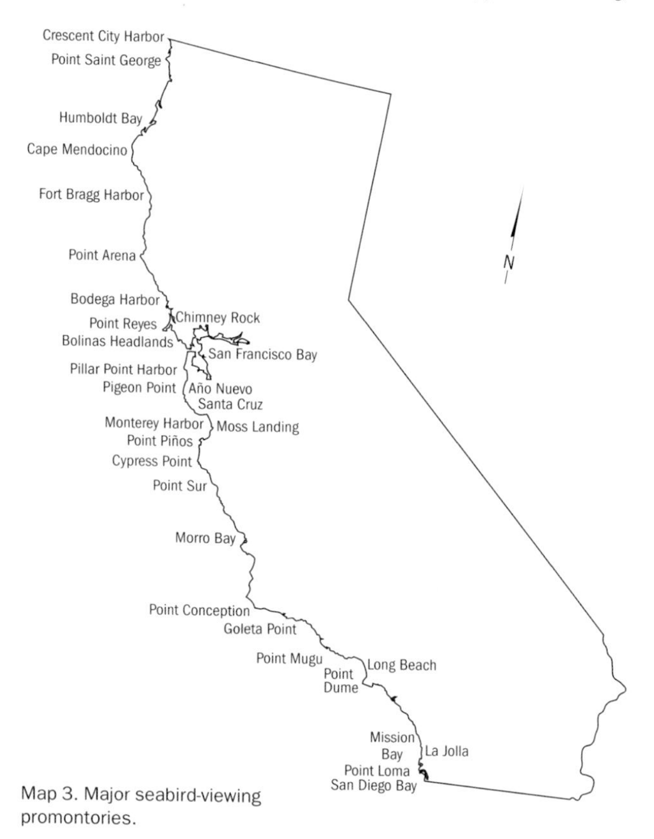

Map 3. Major seabird-viewing promontories.

Plate 27. A Cassin's Auklet crouches in its nest burrow before departing to feed on the nightly upward migration of oceanic zooplankton.

species: species with northern affinities nest mostly in more northerly latitudes and reach their southernmost distribution in California; sedentary species have a breeding range that is centered in California, where they spend most or all of their lives; and species with southerly affinities are distributed mostly in subtropical waters, but their range extends northward into California. These groupings are helpful in understanding the distribution of California's seabird communities during the breeding season, but latitudinal affinities may be quite different in the nonbreeding season. For example, Leach's Storm-Petrel nests in waters from subtropical to subarctic but spends its winter in equatorial waters.

Species with Northern Affinities

Cassin's Auklet, the second most abundant of California's breeding seabirds, is considered subarctic throughout its breeding range and is confined largely to the upwelling domain of the eastern Pacific Ocean. An estimated 36,000 auklets (two-thirds of the California population) nest on the Farallon Islands in such density that only the prosody of Leon Dawson (1923) can give an impression:

> Cassin's Auklets are everywhere on the Southeast Farallon. Burrows, of course, predominate, but there is not a cranny, nook, cleft, crack, aperture, retreat, niche, cave, receptacle, or hidey-hole, from the water's edge to the summit of the light-tower which is not likely to harbor this ubiquitous bird. The interstices

Plate 28. Tufted Puffins spend most of the year on the open sea. In spring, a pair will find a cavity, usually on an island cliff, in which to nest.

> of all stone walls harbor them by the score. Every cavity not definitely occupied by puffin, petrel, or rabbit, is tenanted by an auklet. . . . Heard alone, the Auklet chorus reminds one of a frog-pond in full cry. As one gives attention to an individual performer, however, and seeks to locate him in his burrow, the mystery and strangeness of it grows. . . . Now it is like the squealing of a pig in a distant slaughter pen. We lift our heads and the stock yards are reeling with the prayers and cries of a thousand victims. And now the complaint falls into a cadence, "Let meee out, let meee out, let meee out!" A thousand dolorous voices take up the chorus. The uproar gets upon the nerves. Is this a bird lunatic asylum? Have we stumbled upon an avian mad-house here in the lone Pacific? And are these inmates appealing to the moon, their absent mistress?

Like the auklet, the breeding ranges of Pelagic Cormorant, Common Murre, Pigeon Guillemot, and Tufted Puffin *(Fratercula cirrhata)* also extend from Arctic waters southward into the temperate waters of central California. It is notable that many of Cal-

Plate 29. An omnivorous Western Gull carries off a Cassin's Auklet that failed to reach the safety of its burrow before daybreak.

ifornia's seabirds with more northern affinities (Common Murre, Marbled Murrelet, Pigeon Guillemot, Cassin's Auklet, and Tufted Puffin) are members of the alcid family. Like the penguins of the Southern Hemisphere, alcids have never crossed the effective geographical barrier of the warm tropical oceans to invade the opposite hemisphere, and so, in parallel, each family has evolved in its own quarter of the globe. The physical similarity of penguins and alcids is a classic example of convergent evolution: organisms responding to similar ecological influences with similar morphological and behavioral adaptations. These species have one big difference, however: alcids can fly and penguins cannot.

Sedentary Species

Three taxonomically distinct species—the Ashy Storm-Petrel, Brandt's Cormorant, and Western Gull—remain in the California Current year-round and therefore qualify as "sedentary." Nearly the entire world population of Ashy Storm-Petrels congregates in Monterey Bay each fall in a small patch of deep water. This phenomenon has raised concern about the vulnerability of the species to a single catastrophic event—an oil spill or a toxic outbreak in their food source. Brandt's Cormorant and especially Western Gull are much more widely distributed, and the population of each is apparently on the rise in California. It should be noted that the increase in Western Gull populations is proving hazardous to several other seabirds; Ashy Storm-Petrel and Cassin's Auklet are preyed on heavily by the large and aggressive Western Gull.

Species with Southerly Affinities

Other species are more closely associated with warmer, more southerly waters but move northward after breeding. Brown Pelican, perhaps the most emblematic bird of the California coast, historically bred on all the Channel Islands and as far north as Point Lobos at Monterey. Indeed, anecdotal evidence suggests that Brown Pelicans may even have nested in San Francisco Bay when the earliest Spanish explorers sailed in on their frigates. Alcatraz, the famous island that sits prominently inside the Golden Gate, derives its name from the Portuguese word *alcatraz,* which means both "bucket" and "pelican," a reference to the bird's bucketlike beak.

Another southerly species, the inconspicuous Xantus's Murrelet *(Synthliboramphus hypoleucus),* occurs only between central Baja California and Point Conception, with the bulk of the population centered on Santa Barbara Island and the northernmost colony on San Miguel Island. The birds on Guadalupe Island, Mexico, are a separate subspecies from those that breed in California. Unlike the other alcids, this species has adapted and evolved in the warmer water conditions of the Southern California Bight. These ground-nesting birds are in serious threat of ex-

Plate 30. Heermann's Gulls, which breed on islands off Baja California, are found along the California coast mainly in fall and winter. They often accompany flocks of feeding Brown Pelicans, snapping up fish that escape the pelicans' pouches.

Plate 31. In the fall, postbreeding Elegant Terns move northward along the California coast from their nesting islands off Mexico.

tinction because of depredation by introduced black rats and feral cats. Efforts at predator control have shown some success; but the ultimate fate of this murrelet is still uncertain. After breeding, parents accompany their young out to sea and concentrate in the Southern California Bight, then tend to move northward and scatter along the waters 10 to 60 miles offshore from Point Conception northward to Monterey. Occasionally these postbreeding dispersals bring Xantus's Murrelets northward as far as the Farallon Islands and even to Oregon, but they usually retreat to waters south of Point Conception.

Other southerly nesting species that move northward following breeding include, most commonly, Heermann's Gull *(Larus heermanni)* and Elegant Tern.

Seabird Communities

The seabird community of California includes slightly more than 100 species but is dominated by about 30 species that reach peak numbers in the zone of coastal upwelling. Fall and winter are the seasons of maximum abundance, with an estimate of four to six million birds present in California waters; this number may have been substantially higher prior to the 1870s, before human exploitation disrupted the food web. Sometimes a single species can instantaneously swell the numbers by more than a million individuals, as do Common Murres and Cassin's Auklets during a strong breeding season, or Sooty Shearwaters and Red Phalaropes during migratory pulses. Indeed, in good flight years during the summer months, upward of five million Sooty Shearwaters may arrive en masse, and single flocks of more than one-half million birds have been reported. One of the world's most dramatic natural phenomena—on par with gray whale *(Eschrichtius robustus)* migration and herring runs—is the passage of shearwaters across nearshore waters in summer (southern California) through fall (farther north). From coastal promontories the fortunate observer sees vast rivers of shearwaters streaming by:

> At times the ellipse thus formed extends along the coast as far as one can see, the birds streaming past in close formation . . . gliding close to the water, holding their wings stiffly extended and inclining them so that first one wing then the other just grazes the surface. When the wind is fresh, they soar upward till their forms are outlined against the sky. (Hoffmann 1927)

Aggregations like those shearwaters described by Hoffmann are dramatic, relatively short-lived phenomena. More often, seabirds are widely distributed over an expanse of featureless ocean, featureless at least to human sensibilities. To gain insight into the methods and means by which seabirds negotiate their trackless habitat, scientists and naturalists have paid particular attention to species associations, or assemblages. The behavior of seabirds provides clues to their means of finding and capturing prey. Some species—Common Murre, Brandt's Cormorant, Black-legged Kittiwake, and Western Gull—are particularly adept at finding concentrations of plankton, squid, or fish. As these "finders" aggregate, they may have the effect of concentrating their prey into tighter schools or "clouds," as swarming plankton is sometimes called. Other species then target the behavior of these most astute seabirds and join in the feeding frenzy. Seabird ecologists call the finders "nuclear" or "catalyst" species, and the latecomers are called "joiners." (Seagoing naturalists and fishermen also could be classified as joiners, noticing a feeding flock of gulls circling over an otherwise indistinguishable tract of ocean and rushing into the fray.) As the joiners arrive—Sooty Shearwaters are notorious joiners—the intense hunting pressure they exert may cause the prey to disperse and food concentrations to diminish. Joiners are therefore called "suppressors" by ecologists. This dynamic of aggregation and dispersal is one of nature's rhythms that must have been played out over eons and been one of the driving forces of evolution among the multitudes that spend their lives in the ocean.

An example of the finder–joiner dynamic is readily apparent to the curious observer along California's shore. Brown Pelicans often act as effective finders, cruising along the breaker line with a symmetry that seems to defy aerodynamic logic, given their bulky, prehistoric shape. Then, a lead bird will bank and dive, followed in turn by others in the squadron. Moments later, Heermann's Gulls, attentive to the change in the Brown Pelican's behavior, converge on the roily chaos, stealing fish that the pelicans have dropped off the surface of the water. Such piracy is simply opportunistic foraging, with a *commensal relationship* existing between pelican and gull. Another finder–joiner pairing is more akin to *kleptoparasitism*, a behavior that describes the foraging habits of skuas and jaegers, the "hawks of the sea." When flocks of Elegant Terns are moving up the coast in late summer and fall, they are often shadowed by Parasitic Jaegers; after capturing a fish, the tern may be harassed by a Jaeger until it drops its prey.

Seabird Nesting Sites

All animals and plants are limited by some, or several, environmental variables. The availability of islands, and therefore nesting sites, probably determines the number of seabirds that can nest in California, although during warm-water years, food availability may also be a limiting factor. The Farallon Islands off of San Francisco and the Channel Islands off of Santa Barbara host vast breeding colonies of seabirds—murres, cormorants, auklets, and storm-petrels. Small sea stacks and rocky islets scattered along the coastline also provide refuge, roosting, and nesting sites for small satellite colonies. As mentioned above, the Farallon Islands support the greatest abundance of nesting seabirds on the west coast of North America south of Alaska. With one-half million individuals of 11 seabird species, these islands host the largest breeding populations in the world of Ashy Storm-Petrel, Brandt's Cormorant, and Western Gull and a significant proportion of the world's population of Pigeon Guillemot and Cassin's Auklet.

Second only to the Farallon Islands as important seabird nesting sites, the Channel Islands cluster provides important nesting habitat for many seabirds. Although compared to the California Current, the waters of the Southern California Bight are relatively impoverished foraging grounds, viable seabird nesting colonies exist, especially on San Miguel and its satellite islands—Prince Island and Castle Rock. San Miguel, located 26 miles offshore from Point Conception, is the most northerly and westerly of the Channel Islands and lies at the northern edge of the Southern California Bight. This location places Point Conception fully under the influence of the California Current and ensures that it experiences as much wind, fog, and wave action as other promontories such as Point Reyes, Cape Mendocino, and the Humboldt coast. An estimated 15,000 pairs of seabirds nesting on the San Miguel Island group qualify it as the largest California seabird rookery south of Point Conception, accounting for fully one-third of seabirds nesting in southern California. San Miguel is also significant in the history of west coast ornithology, for it was here that California lost one of its most eloquent and talented naturalists. In 1932, Ralph Hoffmann, Director of the Santa Barbara Museum of Natural History and author of the classic *Birds of the Pacific States* (1927) fell to his death while collecting plants from a cliff face now known as Hoffmann Point. A quote from Hoffmann reveals what turned out to be his fatal enthusiasm: "An ardent bird student will not rest till he has visited some 'bird rocks,' where the perpendicular cliffs are the home in the breeding season of gulls, cormorants, murres, and puffins" (Hoffmann 1927).

West Anacapa and Santa Barbara Islands still support Brown Pelican colonies; however, other breeding sites have been lost or abandoned since the 1960s as a result of eggshell thinning from pesticide contamination. Although the remaining colonies have shown signs of recovery, the long-term health of these populations is in question. The Brown Pelican population at West Anacapa Island, the largest in the state, has ranged from as few as 1,100 nests in the late 1960s to as many as 7,500 nests in the late

Myriad coastal rocks—pinnacles, islets, piedras, blancas, pillars, tombolos, and pedestals—cumulatively compose critical habitat for California's seabirds, providing both safe nesting and roosting sites along the length of the shoreline. In light of that importance, President Bill Clinton created the California Coastal National Monument by proclamation on January 11, 2000. This precious monument, extending some 14 miles offshore and spanning the entire coast from Mexico to the Oregon border, provides foraging and breeding grounds for an estimated 200,000 seabirds. Coupled with the Point Reyes National Seashore, established by John F. Kennedy in 1962, this coastal monument forms a priceless conservation unit. Such enlightened leadership continues to be required in order to stem the degradation of habitat that threatens the health and viability of California's seabird populations. (See "Conservation Concerns and Population Trends," below.)

1980s. The numbers declined in the 1990s as successive El Niños (1990, 1992, 1998) caused a near collapse of the Northern Anchovy *(Engraulis mordax)* stock, the favored prey item of Brown Pelicans during the breeding season. Santa Barbara Island is also an important breeding site for Xantus's Murrelet; however, that population is apparently in decline because of both avian and mammalian predators.

Small offshore rocks and sea stacks, often rather close to shore, and precipitous headlands are also important breeding sites for California's seabirds. Outside the Golden Gate, Seal Rocks and Murre Rock support small colonies of cormorants and murres, and off Cape Mendocino, False Cape Rocks, Sugarloaf Island, and Steamboat Rock support diverse seabird colonies, including a significant proportion of the state's murre population.

Seabird Habitats at Sea

Coastal upwellings, the upwelling frontal zone, and warm, clear, thermally stratified waters of the California Current, along with the shallow nearshore zone, constitute a diverse and dynamic oceanic environment that supports some of the greatest concentrations of seabirds in the world. Let's consider the three broad zones of open water that support different assemblages of seabirds.

Shallow Subtidal Habitat: The Inshore Zone

The waters that overlie the continental shelf, shoreward of the shelf break, are known as the inshore zone. They are generally less than about 650 feet in depth, and seabird densities compare to the highest in the world. The continental shelf is very narrow (3 to 20 miles) along much of northern and central California but broadens to 30 to 45 miles off Eureka, San Francisco, Morro Bay, and southern California. Although there is great seasonal variation, the most common species that frequent these inshore waters include Pacific Loon *(Gavia pacifica)*, Western Grebe *(Aechmophorus occidentalis)*, Brown Pelican, Brandt's and Pelagic Cormorants, Sooty Shearwater, Surf Scoter *(Melanitta perspicillata)*, Common Murre, and Western Gull.

In the shallowest waters closest to shore, Surf Scoter and White-winged Scoter *(Melanitta fusca)* feed on bottom-dwelling *(benthic)* invertebrates, especially small clams, and just beyond the breakers, Common and Red-throated Loons *(Gavia immer* and *G. stellata)*, Horned and Red-necked Grebes *(Podiceps auritus* and *P. grisegena)*, and Marbled Murrelet forage for small fish. Surf Scoter is perhaps most familiar to shore-bound naturalists because, as the name implies, it tends to inhabit waters very close to shore, although large rafts of this species, sometimes intermingled with White-winged Scoter or Black Scoter *(Melanitta nigra)*, congregate to roost in calm, leeward bays.

Farther out in somewhat deeper water, large flocks of fish-eating birds follow large schools of migrating fish—anchovies, sardines, and herring—in their sometimes erratic winter wanderings. Abundances of birds vary as the distribution and abundance of the prey base shift by season, water temperature, current dynamics, and surface conditions. Pacific Loon and Brant's Cormorant are particularly gregarious and tend to inhabit waters farther offshore than the others but are often visible from headlands and beaches. These two species congregate in spectacular flocks in early winter when Pacific Herring *(Clupea pallasii)* enter the larger bays and estuaries to spawn.

Dominant prey for California's seabirds includes anchovy, Sardine *(Sardinops sagax)*, Market Squid *(Loligo opalescens)*, surfperch (Embiotocidae), smelt (Osmeridae), and particularly the young of a variety of locally breeding rockfish (*Sebastes* spp.). These species can occur either inshore or farther out over the shelf break, depending on the habitat characteristics of the substrate, the shifting current convergences, and the all-important water temperature and depth of the thermocline. Biological activity, especially primary productivity, tends to be generated at thermal boundaries, which concentrates species up the food chain.

The Continental Shelf Break

The continental shelf slopes gradually from the shore, and then, at the shelf break, drops abruptly from a depth of about 650 feet to a depth of as much as 6,500 feet. This steep submarine escarpment creates its own currents and its own upwelling pattern, which concentrate abundant prey items for seabirds. Krill, small, free-swimming, pelagic crustaceans—shrimp of the genus *Euphausia*—are the most common prey along the shelf break. So abundant are these animals that biologists refer to them as "clouds of the ocean." Krill clouds (along with enormous swarms of *copepods* and *mysids*) attract a variety of marine animals, especially whales and seabirds.

Because of the variation in width of the continental shelf, the shelf break varies in its distance from shore, but on average it is about 30 miles out, beyond the horizon, farther than can be seen even with binoculars or a powerful spotting scope. Deep submarine canyons intersect the shelf at several locations, bringing exceptionally deep water quite close to shore, most notably at Cape Mendocino and Monterey Bay.

During spring, summer, and early fall, these deeper water habitats are visited by Southern Hemisphere shearwaters, members of the family of oceanic birds inelegantly named "tubenoses" that also includes storm-petrels. "Tubenose" refers to the characteristically shaped, double-barreled, tubelike nostrils fixed atop the bill, an adaptation that aids in excretion of salt. Shearwaters and albatross, with long, narrow wings and aerodynamic torsos, are consummate ocean wanderers. Because some shearwaters breed in Australia and New Zealand, California's summer coincides with their austral winter. Sooty Shearwater is the most numerous tubenose in California waters, by far, but other species such as Buller's and Pink-footed Shearwaters and large but elusive flocks of storm-petrels also forage along the shelf break. The larger albatross tend to be more solitary, although Black-footed Albatross sometimes gather in impressive numbers where food is abundant.

As the shearwaters return to their breeding grounds in the Southern Hemisphere in the fall, another tubenose, the Northern Fulmar *(Fulmarus glacialis)*, arrives from its more northerly breeding grounds in the North Pacific—coastal cliffs on the Kamchatka Peninsula, Wrangel Island, and the Aleutian Islands, or the coasts of Siberia and the Alaska Peninsula. And so, these ecological equivalents—shearwaters and fulmars—come from opposite hemispheres, disparate regions of the globe, to occupy the relatively stable California waters in different seasons. This seasonal partitioning of resources is a common phenomenon in California, a function of its geographic location at midlatitudes and the resulting overlap of climatic influences from both north and south.

The fall (postbreeding) distribution of two closely related seabirds—Cassin's and Rhinoceros Auklets—serve to illustrate the ecological richness of the shelf break and its influence on closely related but evolutionarily distinct species. Cassin's Auklet is planktivorous, feeding almost exclusively on zooplankton, clouds of krill, and copepods. Its fall distribution is closely associated with the shelf, and large numbers scatter or cluster in flocks only a mile or so in width directly over the shelf break; few birds move closer to shore or farther out to sea. In contrast, the Rhinoceros Auklet is piscivorous, exclusively fish-eating, and occupies waters slightly more seaward of the break. But Rhinos do not occur in aggregations as large as those of Cassin's Auklet because Rhinos feed higher up on the food chain, and, thus, their prey is less abundant and more widely distributed.

Other species that are characteristic of the break, if less common, include Black Storm-Petrel *(Oceanodroma melania)*, Pink-footed and Buller's Shearwaters, and several species of transient terns. Yet even some of the most pelagic species—Arctic Tern or Laysan Albatross—tend to concentrate slightly shoreward of the break, over the outer edge of the continental slope, primarily in clear waters just outside the upwelling zone.

Deepwater Habitats: Beyond the Shelf Break

Seaward of the influence of coastal upwelling, beyond the continental shelf break where depths can exceed 6,500 feet, the surface water tends to be relatively clear and warm, and less nutrient rich than above or inside the shelf break. Seabird densities are very sparse in these trackless ocean waters, although it is also the least known of the oceanic habitats and considered "the last frontier" by seagoing naturalists. Here, bird abundance reaches only about 5 percent of the densities found inside the shelf break.

Species that frequent the deepest ocean include graceful, long-distance migrants—Arctic Tern, Sabine's Gull, Black-legged Kittiwake, Long-tailed Jaeger—masterful flyers all. The large surface area of the wings compared to a relatively small body size (called the wing to body aspect ratio) of these peripatetic pelagics allows them to cover vast reaches of barren ocean in their search for schools of surface-feeding fish upon which they prey. Leach's Storm-Petrel is another species that frequents waters seaward of the continental slope, apparently favoring the relatively clear waters on the outer edge of the upwelling zone. Leach's Storm-Petrel nests on sea stacks and rocks nearshore off northernmost California (especially Point Saint George); being nocturnal and pelagic, this small, swallowlike sprite is seldom seen from shore. Although present in California waters year-round, it is most common in summer, and the ranks of the relatively small breeding population (more than 15,000 individuals) are swelled by 10 times that number in midsummer, when birds from elsewhere assemble in California.

Submarine canyons, also called nearshore escarpments, intersect the continental slope in places and so bring deeper water and associated pelagic species closer to land at several locations. Two of these submarine canyons—Monterey and Ascension Canyons—account for the excellent pelagic birding in Monterey Bay and attract oceangoing species such as Black-footed Albatross, Ashy Storm-Petrel, and Xantus's Murrelet close to shore during summer and fall, and Northern Fulmar, Red Phalarope, and Black-legged Kittiwake during winter and spring.

Other members of the tubenose group that look similar to the shearwaters, the *Pterodroma petrels*—Stejneger's Petrel *(P. longirostris)* and Murphy's Petrel—seem averse to landforms and inhabit only the deepest ocean waters; they are found off southern California only beyond the 1,500 fathom (9,000 foot) contour. Deep basins (more than 3,300 feet) and steep escarpments surround the islands within the Southern California Bight, creating complex circulation patterns that provide rich feeding grounds for seabirds. The Santa Barbara Channel attracts many deepwater-loving species such as Buller's and Flesh-footed Shearwaters and also rarer tubenoses such as Streaked Shearwater, which has been sighted there in recent years.

Seabird Behavior and Ecology

Seabirds tend to occur in clusters, congregating at apparently featureless tracts of ocean, and then disperse as suddenly as they arrived. It is obvious to the sea-bound naturalist that these seabird throngs are cueing on prey, but it is not so clear how they locate clouds of krill and zooplankton. Recent research by University of California at Davis scientists found that some species may be keying on the aroma of the gas dimethyl sulfide, given off by microscopic plant plankton eaten by krill. Apparently some species are more sensitive to the aroma than others, and a hypothesis has been developed to explain the discrepancy. Storm-petrels are primarily nighttime foragers and therefore cannot always rely on visual cues to locate prey; perhaps evolution has refined their olfactory skills to aid in nocturnal foraging. Large tubenoses, for example, albatross and shearwaters, are more diurnal in their foraging behavior and may use visual clues such as surfacing marine mammals, flocks of storm-petrels, or current convergence traces to alert them to prey concentrations:

> The migratory flights [of shearwaters and petrels], though not yet properly investigated, appear to be as definite as those of landbirds, and there is little ground in believing that any of the species wander over oceans indiscriminate, though by what the birds recognize particular areas of the ocean and find their way . . . is a complete mystery. (Alexander 1954)

This ability to discern and zero in on the aroma of plant plankton might help shed light on the question of how seabirds manage to navigate across the trackless open ocean with few

Pelagic birding trips, crowded with naturalists shrouded in foul-weather gear, are almost as famous as owling trips for many hours of little or no bird activity. Once, in Monterey Bay about eight miles off Point Pinos, we had been trawling and chumming for several hours with only an occasional jaeger or kittiwake to spark our enthusiasm. Then the boat slowed to allow us a look at a small, dark-bodied petrel that the leader had spotted bobbing in the swell. As we tried to decide whether it was an Ashy or the less common Least Storm-Petrel, we found ourselves suddenly surrounded by hundreds of White-sided Dolphins, breaking the surface, rolling through the swells and troughs, seeming to arrive to this calm patch of ocean instantaneously, from everywhere and nowhere. No sooner had the dolphins appeared than the surface of the water all around rippled with the patter of petrel feet as thousands of Ashy Storm-Petrels appeared, seemingly out of thin air. Soon shouts of "Skua!" "Rhino!" "Flesh-foot!" "Black-footed Albatross!" rang from the lungs of exhilarated birders. Over the next half hour, we saw virtually every species of seabird we could expect on a pelagic trip, including a few relative rarities. Finally, a river of Sooty Shearwaters surrounded the boat; their sheer numbers and frantic antics seemed to overpower all the other species combined. Then, instantaneously, the melee dispersed, a calm returned to the sea, and only a few shearwaters and gulls remained in the wake.

visual reference points. Plankton tends to be concentrated in current convergences and upwelling zones, oceanographic features associated with seafloor topography—shelf breaks, sea mounts, and submerged ridges. The result is a "smellscape" in the air that corresponds to the underwater landscape that promotes the phytoplankton blooms, which in turn generate a gaseous mist. For some seabirds, following their noses might be as good as reading a map. Unlike the tubenoses and the Turkey Vulture *(Cathartes aura)*, in which natural selection has highly refined the olfactory sense, the ability to smell is rudimentary in most birds.

Foraging strategies, or abilities, of seabirds also can be understood by grouping species according to dipping or diving depth. Species that forage essentially on the surface, or dive to very shallow depths of a few yards or less, include pelicans, shearwaters, storm-petrels, gulls and terns, and phalaropes. These surface feeders tend to pursue small, fast-moving or widely dispersed prey items and are capable of searching over large stretches of ocean in their foraging effort. Sooty Shearwater, which searches out krill and squid around the entire Pacific, may be the ultimate example of a surface-feeding seabird. Those species that dive to intermediate depths, to as deep as 260 feet, are the Double-crested Cormorant and Cassin's and Rhinoceros Auklets. At these depths, Double-crested Cormorant is feeding on larger, solitary, bottom-dwelling fish, the Rhino principally on midlevel schooling fish such as anchovies, and Cassin's Auklet primarily on zooplankton (principally krill) and some rockfish. Those species capable of diving to depths of 330 feet or more include Brandt's and Pelagic Cormorants, Common Murre, Pigeon Guillemot, and Tufted Puffin. These, too, partition the resources at depth among the various prey available. When prey is abundant, there may be substantial dietary overlap among several species. For example, in years when rockfish are abundant on the Farallon Islands, both cormorant species and Pigeon Guillemot eat mostly juvenile rockfish. In low-rockfish years, which correspond to warmer water years, their diets are diversified: Pigeon Guillemot may take more octopus and sand dabs, Pelagic Cormorants may take more mysid shrimp, and Brandt's Cormorant may take more flatfish. In a variable environment, the ability to shift foraging behavior as the availability of prey varies grants an important advantage for survival; you would expect successful species to display this capability. So, Common Murre may forage primarily on juvenile rockfish in cool-water periods and then, during warm-water El Niño years when few rockfish are produced, will move closer to shore to exploit anchovies, smelt, sand dabs, and other replacement species.

Some seabirds are generalists, feeding on a wide variety of available prey within their foraging range when conditions require, whereas others specialize on specific prey items. The tendency to forage and feed in flocks, as opposed to solitarily, is also an indication of the behavior and distribution of a prey. Cassin's Auklets gather in fairly dense concentrations to feed on krill swarms, Pigeon Guillemots feed somewhat loosely aggregated, whereas Pelagic Cormorants are mostly solitary feeders, diving among rocks in search of benthic organisms.

Conservation Concerns and Population Trends

Seabirds of the California Current respond differently to environmental challenges and changing conditions depending on their feeding strategies, natural histories, and ability to respond to human influences. California's seabirds display examples of declining, stable, or even increasing populations, and it was probably ever thus. Unfortunately, the increase of human impacts (e.g., oil spills) and what appears to be an increasing frequency of warm-water events may have accelerated or increased the environmental variability with which seabirds have to contend. The result is a decline in the more sensitive and specialized species (e.g., Ashy Storm-Petrel) and an increase in the more aggressive and generalist species (e.g., Western Gull).

Of the most abundant species, the shallow-diving, surface-feeding Sooty Shearwater and Cassin's Auklet declined precipitously from the mid-1980s to the mid-1990s as the result of increases in sea temperatures and thermocline depth. The shearwaters declined by an astounding 90 percent during that decade (Veit et al. 1996), and because they were overwhelmingly the most numerous seabird in the California Current, seabird abundance declined within the region by 40 percent over that period. Because Sooty Shearwater is so mobile and adept at finding suitable habitat, the population probably shifted its migratory routes and frequented waters beyond the California Current.

Common Murre, a dominant member of the breeding seabird community, has had a variable history of decline and recovery, but numbers have suffered in recent times because of mortality associated with the gill-net fishery and oil spills, and low reproductive success caused by the increasing frequency of El Niño events. In central California the population was decimated in the late 1800s and early 1900s by an "egging" operation out of San Francisco. After the chicken poultry industry was established and the demand for Common Murre eggs was reduced, the population rebounded and grew to several hundred thousand breeding pairs by the early 1980s. This recovery, however, was marred by intense mortality associated with drowning by gill netting, with 70,000 birds killed between 1979 and 1987. Legislation to abate "by-killing" was enacted in 1987, and the Common Murre was given a reprieve, only to be then hit by the 1984 Puerto Rican and the 1986 Apex Houston oil spills, among others, and several intense El Niño years. The central California population declined by 60 percent during the 1980s and has not recovered substantially since.

Cassin's Auklet, too, has suffered declines due primarily to increase in predation by Western Gull on its breeding grounds at California's largest colony, the Farallon Islands. The correlation between an increasing Western Gull population and a decreasing Cassin's Auklet population is emblematic of the unintended consequences of human urbanization of California's coast. Gulls on the Farallon Islands commute daily from Bay Area landfills and garbage dumps to the islands (20 miles or more) and so are subsidized by the refuse of the urban culture. Auklet declines have also been reported at Prince Island in the Channel Islands.

Other less common species—Ashy Storm-Petrel, Marbled Murrelet, Xantus's Murrelet—are also suffering from habitat destruction and predation; however, some positive increases are occurring concurrently. Most dramatic has been an increase in Double-crested Cormorants—some of this is attributable to occupancy of artificial nesting sites in San Francisco Bay and reduction in marine pollution in southern California—and Western Gulls, for reasons mentioned above. Rhinoceros Auklets, countering the trend of other alcids, have increased in California. Rhinos reoccupied the Farallon Islands in the early 1970s, after a 100-year absence, and have recently extended their nesting distribution southward to the Channel Islands.

Populations that appear stable, although with expectedly poor reproductive success during El Niño events, are Brown Pelican, Brandt's and Pelagic Cormorants, Fork-tailed Storm-Petrel *(Oceanodroma furcata)*, Pigeon Guillemot, and Tufted Puffin. Tufted Puffins, like Rhinoceros Auklets, have recently extended their range southward into the Channel Islands.

SHORT-TAILED ALBATROSS

The history of the Short-tailed Albatross *(Phoebastria albatrus)* offers a glimmer of hope for seabird populations and for the human ability to reverse transgressions. During the nineteenth century, before its population was nearly exterminated by plume hunters and hunters at nesting colonies in Japan, this species was a common visitor to California waters throughout the year.

> The Chinese fisherman regard these monarchs of the high seas [Short-tailed Albatross] with suspicious awe, feeding them and propitiating them with choice bits, in hope of averting danger and winning good luck in their fishing. According to their belief, the whales drive the sardines into the [Monterey] bay to help the Chinese, but the albatross drives the whales. (Wheelock 1910)

Now it is the rarest of all albatrosses. Although considered extinct for some years in the early twentieth century, a colony of 10 birds was discovered on Torishima (a volcanic island 300 miles south of Japan) in

Plate 32. A Short-tailed Albatross rests on Midway Island.

1953, and subsequently protected. It has rebounded steadily since. The U.S. Fish and Wildlife Service has initiated an innovative approach to attracting this bird to breed on Midway Islands in the central Pacific, where individuals sometimes loaf among the Laysan Albatross colonies. Models of Short-tailed Albatross have been installed together with a speaker system to broadcast their calls. Although the world population is still small, individuals have been reappearing, if sparingly, in California since 1988. May the future welcome these monarchs of the high seas back to our shores with open arms.

BIRDS OF THE SHORELINE

A long finger of the ocean's hand
Explores our coastline,
Follows the seam of the San Andreas,
Drowns a valley,
Makes a pristine bay
Filled with black geese,
Butternoses and bluebills . . .
. . . A bay patrolled by yellowlegs
And willets with semaphoric wings.
Egrets like prayer flags
Flutter along the shoreline . . .

MICHAEL WHITT, "TOMALES BAY" *(TAMAL-LIWA)*

California's shoreline spans nearly 10 degrees of latitude, a distance of about 1,100 miles. This vast and varied coastal boundary provides a diverse mix of habitats—precipitous rocky shores and sea stacks, wave-sculpted beaches and shifting sandbars, dunes and swales, wide river mouths and moist coastal plains, and, most importantly for birds in terms of overall abundance,

Map 4. Shoreline Bioregion.

large embayments containing tidal flats and marshes, brackish lagoons, salt ponds, and small freshwater marshes. These wetlands, especially those with expanses of tidal flats flooded by the tide twice daily, and large areas of open water, are where flocks of waterbirds congregate.

Waterbirds, or aquatic birds, that use coastal wetlands include approximately 120 species from 16 avian families. About half of these species are referred to generically as shorebirds and waterfowl, whereas the others are lumped into a catchall category of "other waterbirds." Some waterbirds, Brown Pelican *(Pelecanus occidentalis)*, for example, may occur in the ocean as well as in coastal wetlands, but are not really "ocean birds" because of their reliance on shallow-water coastal wetlands for much of their foraging and roosting habitat. Waterbirds depend on wetlands during all seasons of their lives as crucial habitat for feeding, roosting, and breeding. Some waterbirds use interior wetlands for one portion of the year and coastal wetlands for another. For example, the elegant, long-necked Western Grebe *(Aechmophorus occidentalis)* breeds on freshwater lakes in the inner Coast Ranges or the Great Basin and then winters on coastal bays.

Although most species are associated with a favored microhabitat within a wetland system, it is important to recognize that birds respond to a mosaic of wetland types and that a large wetland containing open water, tidal flats, salt marsh, and salt ponds

Plate 33. The Willet is a common shorebird. Except for some nonbreeding individuals, most leave the coast in spring for their nesting grounds in Great Basin meadows.

supports more species and more individuals than any single habitat type. When tidal flats are flooded, shorebirds may need to roost or forage in an adjacent tidal marsh, salt pond, moist pasture, or floodplain. When strong winds roil the open water, waterfowl tend to huddle in leeward bays, adjacent ponds, or flooded fields. When winter storms buffet the coast, shorebirds may commute inland, sometimes great distances, to forage in flooded pastures. Various studies have attempted to quantify bird use of adjacent habitats, and invariably they have come to the conclusion that the greater the diversity of available habitat the greater the use by waterbirds. That said, during nonmigratory periods, most

Plate 34. Marbled Godwit, with their long bills, can access creatures deep in the tidal flats, thus utilizing a niche unavailable to shorter-billed shorebirds.

species are relatively sedentary, using the same areas within an estuary or the same sandy shoreline throughout the season. For example, in San Francisco Bay, those common long-legged shorebirds, Willets *(Catoptrophorus semipalmatus)* and Marbled Godwits *(Limosa fedoa)*, forage on the same South Bay tidal flats and roost on the same salt marsh island throughout the winter. Sanderling *(Calidris alba)*—that short-legged, pot-bellied shorebird often seen on beaches chasing outwashing waves—may visit the same stretch of beach during each tidal cycle. Other waterbirds may move among coastal sites under varying conditions. So, American Avocets *(Recurvirostra americana)* or Black-crowned Night-Herons *(Nycticorax nycticorax)* "commute" with some regularity between San Francisco Bay and Bolinas Lagoon, 20 miles up the coast. Other waterbirds are somewhat more prone to longer movements and have large-scale distributional shifts midseason. Large shorebird flocks of Dunlin *(Calidris alpina)* com-

Plate 35. A flock of American Avocets take flight from winter wetlands.

Plate 36. Sanderlings, mainly winter visitors to California, are often seen in small groups scuttling after receding waves in their search for Mole Crabs and other sand-dwelling invertebrates.

mute from San Francisco Bay to the Sacramento Valley after heavy winter rains. Northern Pintail *(Anas acuta)*, a dabbling duck, may arrive in a brackish coastal lagoon in early fall and remain until winter rains create shallow freshwater habitat inland, then move, en masse, to these newly flooded wetlands.

California contains some of the most productive coastal wetlands on the Pacific Coast and supports, especially during migration and winter, a large proportion of the global populations of several waterbird species. Through the rest of this chapter, we discuss these various coastal wetlands and the taxonomic groups of waterbirds that rely on the wetlands' health, complexity, and protection for their continued existence.

A few general patterns of waterbird distribution and abundance are evident in California's coastal wetlands:

- Approximately two-thirds of waterbirds are represented by three families: Anatidae (waterfowl), Scolopacidae (sandpipers), and Laridae (gulls and terns).
- Winter is the season of greatest waterbird abundance and species diversity on the coast, because of the overlapping presence of resident and overwintering populations in this relatively hospitable climate; individual species, however, may peak in numbers during fall and spring migration.
- Tidal flat specialists (sandpipers) rely on a network of coastal wetlands for their migration and wintering sustenance and benefit from habitat diversity within a given wetland system.
- Abundance and density of more-marine species such as Brown Pelican, Surf Scoter *(Melanitta perspicillata),* and Common Murre *(Uria aalge)* is limited almost exclusively to the immediate coast.
- Some species of diving ducks such as Greater Scaup *(Aythya marila),* Bufflehead *(Bucephala albeola),* and Ruddy Duck *(Oxyura jamaicensis)* tend to occur more abundantly in coastal wetlands than in interior wetlands; others, such as Redhead *(Aythya americana),* Ring-necked Duck *(A. collaris),* and Common Merganser *(Mergus merganser)* are more abundant inland.
- Abundance and density of surface-feeding ducks, such as Northern Pintail and Mallard *(Anas platyrhynchos),* tend to increase with distance from the immediate coast (see "Birds of the Central Valley and Delta"), although impressive concentrations of some dabbling ducks can occur, especially in early winter, at coastal sites.
- More than half of all waterbird species that occur in coastal wetlands are either uncommon or rare in California.
- Breeding habitat for several rare, threatened, or endangered species is provided by coastal wetlands.
- Several of California's wetlands provide habitat of "hemispheric" importance to the world's shorebird populations and are therefore included in the Western Hemisphere Shorebird Reserve Network (see table 1). Sites important to shorebirds are generally important to other waterbirds as well.

TABLE 1. California Coastal Wetlands of Particular Importance to Shorebirds

Wetland	County
Smith River mouth	Del Norte
Lake Talawa and Lake Earl	Del Norte
Humboldt Bay	Humboldt
Garcia River mouth	Mendocino
Bodega Harbor	Sonoma
Estero Americano	Sonoma/Marin
Tomales Bay	Marin
Abbotts Lagoon	Marin
Limantour and Drakes Estero	Marin
Bolinas Lagoon	Marin
San Francisco Bay	Nine bay area counties
Pajaro River mouth	Santa Cruz/Monterey
Elkhorn Slough	Monterey
Salinas River mouth	Monterey
Morro Bay	San Luis Obispo
Santa Maria River mouth	San Luis Obispo
Devereux and Goleta Slough	Santa Barbara
Mugu Lagoon	Ventura
Los Angeles River mouth	Los Angeles
San Gabriel River mouth	Los Angeles
Seal Beach National Wildlife Refuge	Orange
Bolsa Chica Lagoon	Orange
Upper Newport Bay	Orange
Santa Margarita River estuary	San Diego
Batiquitos Lagoon	San Diego
San Elijo Lagoon	San Diego
Mission Bay	San Diego
San Diego Bay	San Diego
Tijuana River estuary	San Diego

Locations in **bold** are included in the Western Hemisphere Shorebird Reserve Network, habitats known to support a minimum of 20,000 shorebirds annually. Numerous smaller wetlands on the immediate coast are also important habitats, but too numerous to list here. However, some of the famously "birdy" spots include Big and Stone Lagoons, Humboldt County; Mattole River mouth, Mendocino County; Russian River mouth, Sonoma County; Rodeo Lagoon, Marin County; Lake Merced, in San Francisco; Pescadero Marsh, San Mateo County; Santa Ynez River mouth and Sandyland Slough, Santa Barbara County; Santa Clara River estuary, Ventura County; Malibu Lagoon and Ballona Wetlands, Los Angeles County; and San Diego River mouth, San Diego County.

Shorebirds

The diverse avian order Charadriiformes includes several quite diVerent types of waterbirds—shorebirds, skuas, gulls, terns, skimmers, and auks. The majority of the "shorebirds" belong to one of two families, the plovers (Charadriidae) or the sandpipers (Scolopacidae). The plovers and sandpipers are similar in appearance; both tend to be small- to medium-sized with longish legs. Plovers are generally stouter bodied with shorter and stubbier bills than the sandpipers, which tend to have longer legs and pointier bills. Oystercatchers (family Haematopodiae) and avocets and stilts (family Recurvirostridae) are also "shorebirds," although more distantly related to the others.

The Pacific Coast of North America is a major flyway, along which shorebirds breeding in eastern Siberia and Alaska migrate to wintering areas from British Columbia to Tierra del Fuego at the southern tip of South America. California's shoreline comprises a significant portion of that flyway and provides productive foraging grounds for both migrating and overwintering flocks of shorebirds that breed in the Arctic. Traveling as they do from above the Arctic circle to the shores of southernmost South America, the "windbirds" are ambassadors of the globe. American Golden-Plover *(Pluvialis dominica)* and Pectoral Sandpiper *(Calidris melanotos)*, among the most peripatetic of this wandering family, may fly 20,000 miles each year in pursuit of an endless summer and a reliable supply of food.

Of the 20 most common shorebird species encountered in coastal California, about two-thirds breed in the Arctic and about one-third breed in the temperate zone. Of temperate-zone breeders, most nest in the interior—some as close as the Great Basin (Willet), others as far as the Canadian prairie provinces (Marbled Godwit). Like their northern cousins, they too abandon their breeding grounds in the interior for the more temperate coast during the nonbreeding season. Shorebirds use the coastal wetlands, beaches, and plains of California intensely as wintering grounds. "Wintering," however, is a misleading modifier, considering that shorebirds are present nearly year-round. The majority breed at the higher latitudes, and their breeding season is compressed into a brief six- or seven-week period from late May to early July. Also, some larger species such as curlews

(*Numenius* spp.) and godwits (*Limosa* spp.) do not return to nesting grounds in their first year of life. After fledging *precocial* young that can fend for themselves, shorebirds waste no time leaving the breeding grounds for fairer climes. Failed or unmated breeders may leave even earlier. Early to mid-June is the only time when few migratory shorebirds are found in California's coastal wetlands.

Arctic-zone breeders that commonly winter along the California coast include Black-bellied Plover *(Pluvialis squatarola)*, Semipalmated Plover *(Charadrius semipalmatus)*, Whimbrel *(Numenius phaeopus)*, Ruddy and Black Turnstones *(Arenaria interpres* and *A. melanocephala)*, dowitchers (*Limnodromus* spp.), and most abundantly, the *Calidris* sandpipers—Sanderling, Dunlin, and Western and Least Sandpipers *(C. mauri* and *C. minutilla)*. Temperate-zone breeders, in contrast, tend to winter closer to their breeding grounds. Although some of those that breed in coastal locations—Killdeer *(Charadrius vociferus)*, American Avocet, Black-necked Stilt *(Himantopus mexicanus)*—move within or between wetland complexes, they tend to stay fairly close to their breeding grounds year-round. Some of the most conspicuous of the interior temperate-zone breeders—Willet, Marbled Godwit, Long-billed Curlew *(Numenius americanus)*—arrive on the coast shortly after the breeding season.

Plate 37. Black Turnstones frequent rocky beaches, where they can be seen flipping pebbles and seaweed debris in their search for food.

Plate 38. The Least Sandpiper has yellowish legs that help to distinguish it from other common "peeps."

Snowy Plover *(Charadrius alexandrinus)*, one of California's more threatened breeding species, nests in scattered pairs and loose groups, mostly along the beaches, dunes, and dry salt pans of the coastal wetland systems. The entrepeneurship of this unassuming little shorebird is revealed in its habit of breeding at one coastal location and then moving to breed a second time at a location hundreds of miles away (Stenzel et al. 1994). Traditional wintering spots include large tracts of relatively undisturbed beaches, river mouths, and sandspits along the immediate coast.

Several locally breeding shorebirds rarely wander far from their breeding territories. Black Oystercatcher *(Haematopus bachmani)* is perhaps the most sedentary of California's shorebirds. Rarely does it venture outside a short stretch of rocky shoreline, not even into coastal lagoons or estuaries immediately adjacent to the sea stack or rocky jetty on which it nests and feeds. Most oystercatchers seem to need the overwashing swell and constant crash of waves against the rocks to feel at home. Not only is the oystercatcher uncharacteristic of a shorebird in behavior, it is also unique in its looks. It sports the most massive bill of all California shorebirds, awl shaped and bright red, a necessary tool for prying or cracking open the tough shells of their favored prey—Owl Limpets *(Ottia gigantea)* and mussels.

Plate 39. A Black-crowned Night-Heron protectively covers its young. Night-Herons nest in colonies, often in company with egrets.

Taxonomically distinct from shorebirds, the long-legged waders, members of the family Ardeidae—especially Great Blue Herons *(Ardea herodias)* and Great and Snowy Egrets *(A. alba* and *Egretta thula)*—are a familiar sight among all habitats within coastal wetlands. Although not restricted to the coast (ardeids are common in interior wetlands as well), they tend to be widely distributed and relatively common members of most coastal wetland communities. Some wetlands support particularly important breeding colonies of these colonial nesters, espe-

Plate 40. The Black Oystercatcher blends so well into its rocky shoreline habitat that its presence is often revealed only by a strident call or a view of its bright red bill.

cially those with predator-free islands. Indian and Teal Islands in Humboldt Bay host the state's largest colony of Black-crowned Night-Heron. Terminal Island in Long Beach is another important site. Ardeids often nest in colonies composed of mixed species; in Artesian Slough in the South Bay, five species nest in dense stands of bulrush.

Other Waterbirds

The coastal embayments, protected from Pacific swells and winter storms, offer refuge to wintering waterbirds that prefer open-water habitats—loons and grebes, pelicans and cormorants, diving ducks and dabbling ducks.

Diving birds in California are a taxonomically diverse group including loons (five species), grebes (six species), pelicans (two species), cormorants (three species), and diving ducks (a dozen common species). The diving ducks, most of which dive to the bottom to forage on mollusks, other benthic invertebrates, and fish, have names as exotic as the shorebirds—Bufflehead, Canvasback *(Aythya valisineria)*, goldeneye, scoter, scaup, and merganser. Diving ducks generally occur in open waters, most often shallow-bay habitat at depths up to about 20 feet, but some species such as Greater Scaup and Surf Scoter can dive considerably deeper. Divers also frequent the edges of human-made structures such as pilings and riprap. Those structures often provide surfaces that support or attract prey—small fish, mussels, barnacles, and a variety of crustaceans. With the exception of cormorants, divers tend to occur most frequently in coastal wetlands during the nonbreeding season, when they may be present in large flocks or "rafts." When strong winds buffet the coast, divers tend to "raft-up" in protected leeward coves.

For most of the divers, California's coastal wetlands are critical wintering grounds. Greater Scaup and Surf Scoters comprise the majority of diving ducks on shallow coastal estuaries, but other species are well represented at some specific sites, each of which has its own unique attributes. Years of surveys from several coastal sites, most notably by Point Reyes Bird Observatory and the U.S. Fish and Wildlife Service, have quantified the value of coastal wetlands to divers. San Francisco Bay, the dominant coastal wetland at the center of the state, was found to support 56 to 92 percent of the scaup, 33 to 38 percent of the scoters, 46 to

54 percent of Canvasbacks, and 13 to 29 percent of Ruddy Ducks estimated on the entire Pacific Flyway in midwinter surveys by the Fish And Wildlife Service. Midwinter surveys of the bay in the early 1990s estimated a midwinter population of 700,000 waterfowl, mostly diving ducks; this number, although impressive, represents a 25 percent decline since surveys began in the 1950s. On Tomales Bay, a relatively small inlet in Marin County, three species—Surf Scoter, Bufflehead, and Greater Scaup—accounted for 70 percent of the total number of waterbirds present in winter, which averaged over 20,000 individuals annually. Tomales Bay differs from the San Francisco Bay in that it supports higher proportions of Bufflehead and Brant *(Branta bernicla).*

Dabbling ducks, or surface-feeding waterfowl, commonly called "dabblers," generally occur in shallower water than divers and some, such as Green-winged Teal *(Anas crecca),* may forage on exposed tidal flats. Others like Northern Pintail and American Wigeon *(A. americana)* tend to congregate in brackish lagoons at the mouth of freshwater feeder streams. Dabblers feed mostly on vegetation or very small invertebrates. Surface-feeding ducks are mostly vegetarians, and each species has a unique method of foraging by siphoning plant material off the surface. Some species, such as Mallard and Cinnamon Teal *(A. cyanoptera),* breed in coastal tidal marsh or even upland vegetation around the wetland margin. Other common dabblers, most notably Northern Shoveler *(A. clypeata)* and Green-winged Teal, breed only sparsely in California, but winter in appreciable numbers. Although they are important members of the avian community in coastal wetlands, dabbling ducks are even more common in Central Valley wildlife refuges and are treated more fully in "Birds of the Central Valley and Delta." The most common dabblers at coastal sites are American Wigeon, Northern Shoveler, Northern Pintail, Gadwall *(A. strepera),* and Mallard.

Geese do not fit easily into the diver or dabbler categories; perhaps "grazers" is the best description, particularly for the ubiquitous Canada Goose *(Branta canadensis),* which has increased so dramatically along the coast in recent decades, adapted as it is to golf courses, flooded fields, and human-made ponds. Other geese are not particularly well represented on the coast, except for the splendid "sea goose," the Brant. A medium-sized goose with a snazzy white collar, Brant can occur in vast flocks at sites where open water is protected enough and clean

Plate 41. Sometimes overlooked because of its similarity to the female Mallard, the Gadwall is a dabbling duck, "tipping-up" from the surface of the water to feed.

enough to support submerged eelgrass *(Zostera maritima)* beds, on which Brant graze during migration. Arctic breeders, Brant migrate usually well offshore in their fall passage southward, but when they move northward in spring, numbers build in the bays, especially in March and April. Important foraging grounds for Brant include Humboldt Bay (where 30,000 birds can congregate), Tomales Bay and the Point Reyes estuaries, and Morro, Mission, and San Diego Bays. Although formerly a winter visitant, Brant shifted its distribution southward in the mid-1960s, probably because of hunting pressure or eelgrass depletion in California, and has only recently begun to replenish its winter presence here. Thousands still pause at coastal estuaries—especially San Francisco, Tomales, and Humboldt Bays—to refuel during the spring migration northward, however.

Suisun Marsh, in the fresher, upstream reaches of San Francisco Bay, hosts a wintering population of the rare Tule Greater White-fronted Goose *(Anser albifroms elgasi)*, as well as flocks of Snow and Ross's Geese *(Chen caerulescens* and *C. rossii)*. Suisun, however, represents a transition zone between the coast and the Central Valley (see "Birds of the Central Valley and Delta"), and does not really qualify as a "coastal" wetland.

Plate 42. Brant seek out coastal lagoons that are rich in their favorite food, eelgrass. Like marine birds and other sea geese, the Brant has glands that enable it to discharge excess salt through its bill.

Gulls and terns, "the larids," are ecologically and taxonomically related to one another, and tend to congregate in flocks on tidal flats, jetties, islands, pilings, or seawalls. They forage over open water or tidal flats, feeding on a variety of fish that occupy the upper water column. The most common species along the California coast are Western Gull *(Larus occidentalis),* California Gull *(L. californicus),* and Forster's Tern *(Sterna forsteri).* Some species can appear in huge flocks intermittently during migration, for example, Bonaparte's Gull *(L. philadelphia),* or anchovy runs, for example, Elegant Tern *(S. elegans).* Many breed at large coastal estuaries and isolated islands and sandspits, even within the urbanized confines of the San Diego and San Francisco Bays.

Coastal Waterbird Habitats

Rocky Shore

Not all shorebirds are cosmopolitan world travelers, nor are all tidal flat specialists. Black Oystercatcher (discussed above), a species confined to California's most dramatic and rugged

Plate 43. The Wandering Tattler is a coastal nomad frequenting the outer reaches of rocky shorelines. Tattlers breed in the Arctic, usually along mountain streams.

stretches of coastline, tends to stay put, occupying territory on the same few acres of coastline and establishing long-term pair bonds, often for life—as long as 20 years. Although somewhat more migratory, other shorebirds of the rocky coast, such as Black Turnstone, rarely wander beyond the Pacific shore, but find enough sustenance picking limpets, barnacles, and amphipods from craggy crevices to forego a long and arduous intercontinental migration. Like the other rocky shorebirds, Wandering Tattler *(Heteroscelus incanus)* breeds in the Alaskan foothills and winters along the Pacific shore, but, as its name implies, ventures somewhat farther afield, even making oceanic passage to Hawaii or the Galápagos Islands. At each locale, no matter how far flung, it is rocky shore that attracts the tattler.

Four rocky shore species—Black Turnstone, Wandering Tattler, Black Oystercatcher, and Surfbird *(Aphriza virgata)*—may be spotted through a telescope by the observant birder scanning the sea stacks off Big Sur, Bodega Head, or Cape Mendocino, the granitic coast of the Monterey Peninsula, or the riprapped jetty at Princeton Harbor. Of these rocky shore birds, only the oystercatcher remains year-round. As unique as the oystercatcher is, the Surfbird is perhaps the most enigmatic. Most shorebirds have a fairly broad distribution across continents, or even around the northern hemisphere ("circumpolar"), but Surfbirds breed only in the mountains of Alaska and westernmost Canada, far from the sound of surf. This very restricted breeding area is oddly in contrast to the Surfbird's wintering distribution, a narrow stretch of rocky ocean shore stretching from British Columbia all the way to southernmost South America. In California, Surfbirds are very local, occurring rather commonly at some sites, but very sparingly, if at all, at others. It is always a surprise to see Surfbirds at Point Reyes, for example, but they can be expected at Bodega Head, in similar habitat, just 10 miles north. In southern California, they are very rare away from a few reliable localities. Like oystercatchers and tattlers, the Surfbirds' affinity for the rocky shore causes them to avoid tidal flats except in migration.

Another anomalous rocky shorebird is the Rock Sandpiper *(Calidris ptilocnemis)*. Essentially a sedentary resident of Bering Sea shores, some of its population moves southward into northern California. It is seen only rarely along Humboldt's craggier shoreline, and is even scarcer farther south. The apparent scarcity of both Surfbirds and Rock Sandpipers in California in recent decades may be the result of a northward contrac-

Plate 44. The Surfbird is to be found on wave-washed promontories along the Pacific Coast, but it breeds in the interior mountains and alpine tundra of Alaska.

tion of their ranges in winter rather than any population decline, however.

The habitat barrier between the craggy coast and the tidal flat, so evident to the other birds of the rocky shore, must be invisible to both Black and Ruddy Turnstones. Turnstones occur as commonly in bays as on the outer coast, wherever sea grasses, shore-cast wrack, or small pebbles provide habitat for the *amphipods* that turnstones eat. Both species have similar methods of foraging, using their stout bills to turn over pebbles and algae, diligently picking prey from the rubble and wrack.

Ruddy Turnstone, although generally less common than Black Turnstone in California, is in fact a world citizen, distributed perhaps more widely than any other shorebird, certainly more widely than any other rocky shore bird. As a species, Ruddies are as peripatetic as oystercatchers are sedentary. A description of their nonbreeding habitat preferences is instructive: "mostly on and near intertidal zone: rocks, reef, mussel beds; areas of rubble, sand, or gravel; flats where the mud is hard; on lawns on Midway I., and in Hawaii to 6000 ft. elevation on volcanic lava . . . and comes aboard ships at sea. A nearly pelagic shorebird" (Matthiessen et al. 1967).

Rocky shore habitat provides breeding sites for several seabirds, but is perhaps too perilous for any but the most intrepid diving birds. Scoters tend to forage there, and Black Scoter *(Melanitta nigra)* seems to prefer abrupt rocky shorelines to softer bottoms. Harlequin Duck *(Histrionicus histrionicus),* although increasingly scarce along the California coastline, can sometimes be spotted here also, especially along the more northerly portions of the coast.

Sandy Beach and Coastal Strand

Snowy Plovers nest in single pairs, or sometimes in loose colonies, most often on "sandspits, dune-backed beaches, bare beach strands, and the beaches and open areas around river mouths and estuaries. Less commonly, they choose levees of salt ponds, and the floor of dry salt panes for their nests" (Page and Stenzel 1981). Since 1996, small numbers have been found nesting on sand- and gravelbars on the Eel River, seven miles inland from the nearest beach. The sandy shore is a dangerous and difficult habitat for nesting shorebirds. Residential development of coastal dunes and strand at places such as Pajaro Dunes in Santa Cruz County, Stinson Beach in Marin County, and most of the southern California coast has placed homes, people, and dogs in a neighborhood once populated by plovers. The foot traffic of these new residents increases the likelihood that plover nests will be stepped on, or that the birds will be flushed repeatedly, causing desertion of the nest site. With increasing disturbance, the nests are prone to predation by patrolling ravens, perhaps the most devastating predators of plovers and their young.

The discussion of "peeps" (a generic term for small, similar-looking sandpipers) considers three members of the genus *Calidris*—Least and Western Sandpipers and Dunlin (also see "Estuaries and Tidal Flats," below). A fourth species of *Calidris*, the Sanderling, is distinct enough in appearance and behavior not to be lumped with his brethren. Sanderling is a chunky, pot-bellied shorebird of the outer beach, familiar to California beachgoers as it congregates in small flocks and is seen chasing the backwash of a broken wave to forage quickly in the wet sand, and then scurrying back before the next breaker washes over. Like the peeps, Sanderlings prey on invertebrates that burrow in the sand, but they specialize on a larger crustacean, the Mole Crab (*Emerita analoga*), which migrates up and down the beach in that narrow wash zone of breakers. Sanderling behavior has been closely observed at Bodega Bay, and those classic studies shed light on the nature of flocking behavior, territoriality, and foraging ecology in general. J. P. Myers and Peter Connors found that on sections of beach with very low numbers of Mole Crabs, Sanderlings did not defend feeding territories, nor did they defend territories on stretches of beach with very high abundances of prey. Where prey abundance was moderate, however, Sanderlings engaged in intense defensive behavior. So, flocking behavior, a cooperative means of finding prey and a necessary defense against predation ("safety in numbers"), is overcome by competition when prey availability is within certain moderate thresholds.

Estuaries and Tidal Flats

Shorebirds occur wherever the tide ebbs and flows, often in frenetic flocks that flush and land in synchrony, seeming to move with the gracefulness and coordination of a single organism. Aloft, these flocks perform aerial arabesques, turning and

SHORE WATCH

Plate 45. The epitome of aerial predators, the Peregrine Falcon was once compromised by DDT poisoning. An active management program has successfully reestablished it as a breeding bird along the California coast.

The tide is rising, but the mud flats are still exposed and the winter sun shimmers off of them, miragelike. The heat waves are magnified by my telescope, and the air appears viscous. A flock of shorebirds—plovers, godwits, and peeps—feed at the water's edge, but they seem nervous, chattering among themselves and huddled together a little more tightly than usual. Every half minute or so the flock jumps into the air and lands a few yards up or down the shore, then slowly regroups into a tighter flock, drawn together by the gravity of fear.

Suspecting that the flock's jitters are caused by some imminent danger, I scan the adjacent salt marsh in search of a raptor. There, toward the sun, perched on a distant stake, perhaps 300 yards beyond the shorebirds, is a familiar silhouette—broad shoulders, thickset chest tapering through the waist and wing—a Peregrine Falcon. Even at this distance his body language tells me that he is studying the flock. His shoulders are hunkered forward, his head is quarter-cocked to shade his gaze from the strong sun, his attention is focused toward the shore. I doubt the shorebirds can see him, but their nervousness signals that they are aware of his presence.

Suddenly the falcon launches from his perch and courses straightaway across the marsh, contouring low, maybe a foot above the ground. As he approaches the flock, the quiet chit-chat explodes into strident squalls and the flock scatters like birdshot from the barrel of a 12-gauge as the falcon bores through its midst. He banks straight up and climbs several hundred feet, then sails off over the bay waters, flushing teal and scaup with his shadow. I lose sight of the falcon against the boundless, cloudless sky, and the shorebirds settle down to their nervous feeding, again huddling close together.

Moments later, the falcon stoops again, this time approaching out of the blue in a swift down-angled dive, splitting the flock in two. Breaking his fall inches above the mud, he banks upward again, only to loop back down toward where half the former flock has landed. He repeats this loop-the-loop maneuver three or four times in rapid succession before finally selecting a victim. On a final swoop, he reaches out with a single talon and deftly snatches a shorebird from the edge of the now chaotic flock, draws it close into his chest, and flies back toward his roost, low and slow, decapitating the prey on the wing.

The Peregrine is the most skilled of nature's winged predators. Its tapered shape, its long pointed wings, and the purposeful stroke of its flight make up the ultimate flying machine. Peregrines have a peculiar aversion to taking their prey off the ground. They prefer to attack on the wing and usually flush a bird before striking it. That they feed commonly on the "windbirds," or shorebirds—themselves some of the most expert of fliers—and even occasionally on swifts or bats— is a testament to their agility and their speed on the wing. Mock attacks, during which the falcon will stoop on its potential prey in a half-hearted way, are common. These passes may function as test flights, gauging the behavior of the flock and sizing up individual reactions, a way of selecting prey from the masses. Ordinary air speed for a Peregrine is probably in the 25 to 40 mile per hour range, but their stoop speed has been timed at 170 miles per hour on a 30-degree angle of descent and at 220 miles per hour on a 45-degree angle of descent.

Peregrines spend winters along these shores, hunting the shorebird-rich estuaries. In spring they retreat to breed in aeries on secluded mountain cliffs, or on cliffs along the north coast, or even on city high-rises or bridges. In recent years, their numbers have been increasing, in part because of the release into the wild of birds raised in captivity. These captive-breeding programs are an attempt to offset the population decline the birds suffered so severely over the past several decades when DDT and other contaminants filtered through the watersheds of the west, entered the flesh of the windbirds, and became concentrated in the ovaries of Peregrines.

---

swirling, strobing dark and light as they pirouette, disappearing and appearing in an instant—a movement sublime. These choreographed flocks are most often comprised of "peeps," a confederation of difficult-to-distinguish species of small shorebirds that are quite common along the California shore. Here, the peeps include Dunlin (also known as the "Red-backed Sandpiper"), Least Sandpiper, and Western Sandpiper. All arrive here from their Arctic breeding grounds in fairly drab winter ("basic") plumage, often in large numbers. Flocks of 10,000 Western Sandpipers, spread out across a few acres of tidal flat in San Francisco or San Diego Bay in October or November, invariably contain a few hundred Least Sandpipers and Dunlin, all feeding frantically. In a three-year study of San Francisco Bay, the most important wintering area for shorebirds in the western United States, these three peeps accounted for 75 percent of all the shorebirds. In the same study, during spring when migratory peaks swell the numbers even higher, peeps accounted for 88 percent of all shorebirds!

This flocking behavior, so characteristic of the peeps in particular, is a balancing act between exploitation and partitioning of resources—the worms and other small invertebrates that compose their prey—and the safety of the flock from marauding raptors. By November, when the migratory flocks of peeps have mostly settled down for the winter, a predator that specializes in hunting these flocks has also moved in. The Merlin *(Falco columbarius)* is a compact and extremely agile raptor, similar to the Peregrine Falcon *(F. peregrinus)*, but smaller. Merlins specialize in shorebirds, particularly Dunlin. Like the Peregrine, the Merlin surveys the flock from a distant perch, paying particular attention to the individuals that loll on the outskirts or wander slightly farther afield. The Merlin may spend the better part of the winter working the tidal flats in a particular portion of an estuary, and the flock always knows that the raptor is present and ready to strafe the flats at any moment. So they forage frantically, negotiating that fine balance between sustenance and survival.

Perched also in the distance may be that preeminent predator, the Peregrine Falcon, which also preys on the gathered hordes of shorebirds, though Peregrines tend to focus on the larger species—dowitchers, godwits, Black-bellied Plover—or one of the smaller ducks that dabble just offshore—perhaps a Green-winged Teal or wigeon. In fact, older books refer to the Peregrine as the "duck hawk."

The "scolopacid waders" (sandpipers) vary most in the relative length and shape of their bills—from long and curved (Long-billed Curlew) to short and straight (Least Sandpiper). For most of these shorebirds foraging on tidal flats, the ability to obtain prey depends on their own bill length, the depth of the water covering the flats, and the burrowing depth of bottom-dwelling invertebrates. Thus, bill shape is instructive as an indication of the foraging ability and habitat use of each species. The apparent confederacy of the peeps together on the same tidal flats raises the question of competition that is summarized by the ecological concept known as "Gause's exclusion principle" or "Grinnell's axiom," which states that several closely related species cannot coexist in the same ecological niche. To the novice's eye, these flocks of peeps that forage nervously along the tide's edge seem randomly distributed, with no discernable pattern other than that the flock is strewn along the wet sand near the water's edge. But on closer inspection, it becomes evident that the peeps are somewhat segregated, each species cueing on a microhabitat not readily apparent on a human scale. Dunlin, at less than two ounces the largest of the three species and with the longest bill, tends to wade in the shallow water, probing the sand or mud that is slightly flooded. The Western Sandpiper, about one ounce and with an intermediate bill length, is slightly higher up, immediately at the tide's edge. The Least Sandpiper, about three-quarters of an ounce and with the shortest bill, is higher still, often on drier mud. There may be overlap, of course, but in general these microhabitat associations illustrate the narrow niches that these three species occupy.

What are the peeps eating at the tide's edge? Several studies have found that although peeps eat a variety of small invertebrates (amphipods, insects, polychaete worms), tube-dwelling amphipods (*Corophium* spp. and *Allorchestes* spp.) dominate the diet of the Western Sandpiper, the most abundant peep. Indeed, one study at Bolinas Lagoon found amphipods the most important prey of all species of shorebird except Long-billed Curlew. (Curlews specialize on mud crabs and ghost shrimp.) The ubiquity of amphipods as a food item among the dozen or so shorebirds that frequent California's tidal flats suggests that the birds are partitioning the habitat based not on the distribution of prey as much as on their ability to extract it from the substrate. So, among the peeps, all of which forage on soft, moist, sandy-muddy substrate, Dunlin (average bill length about 1.3 inches) is able to probe more deeply than a Western Sandpiper (average bill length about one inch), which in turn probes more deeply than a Least Sandpiper (average bill length about 0.7 inch).

Western Sandpiper is the most abundant shorebird in California, and it has been estimated that two-thirds of the world population winters around the shores of San Francisco Bay. The biogeography of Western Sandpiper exemplifies the distributional versatility of the scolopacid waders in general. Westerns breed mostly in a very narrow band around the Bering Sea and spend their nonbreeding seasons southward along both coasts of North America, on the Pacific Coast from Washington state to northern Peru.

On the California coast, the wintering waterfowl favor mostly leeward coves and tidal shallows, often at the mouths of freshwater streams, where the stream delta provides safe high-tide roost and ready access to the water. Also favored are quiet ponds and marshes with some open water—Elkhorn Slough (Monterey County), and Pescadero (San Mateo County), Bolinas (Marin County), and Big (Humboldt County) Lagoons. At each locale the same species can be found, mixed together in apparent harmony—American Wigeon, Northern Shoveler, Northern Pintail, Green-winged Teal, Gadwall, Mallard—all dabbling ducks. Diving ducks can also be found in these embayments, though they also visit the deeper bays and backwaters. Two species of scaup—Lesser *(Aythya affinis)* and Greater—are difficult to distinguish, but both can be found reliably in calm coastal bays, sometimes roosting in large rafts composed of only their brethren. Numbers of scaup can be impressive. In San Francisco Bay, which holds the bulk of the Pacific Coast population, upward of 200,000 scaup find refuge from October well into spring with numbers peaking in midwinter. Other divers—Canvasback, Ruddy Duck, Bufflehead—are also common, and, like scaup, their numbers vary widely from year to year depending on conditions on the breeding grounds, weather, hunting pressure, and other factors. Less common, especially on the coastal estuaries, are Redhead, Barrow's Goldeneye *(Bucephala islandica)*, Black Scoter, and Long-tailed Duck *(Clangula hyemalis)*. Redheads and Barrow's Goldeneyes are more common inland, whereas the others become more common northward along the coast.

Tidal Marsh

Tidal marsh is vegetated wetland that is subject to tidal action. This habitat has been the most destroyed or altered native habitat type since European settlement of California. In San Francisco

Bay alone, there used to be 190,000 acres of tidal marsh; only 40,000 acres are left, a reduction of 80 percent, and that which remains has been altered drastically. In addition to loss of tidal marsh area, the zone of fringing *halophytes* that forms a natural vegetative transition between marsh and upland has been especially hard hit. This transition zone provides refuge for tidal marsh–dependent birds (and mammals) during high tides and flooding episodes. Diking, livestock grazing, conversion to salt ponds, and urbanization have eliminated or hardened the edges of tidal marshes and have exposed rails and other tidal marsh inhabitants to heavy predation by herons, egrets, raptors, and mammals during high tides. Other pressures that threaten to degrade or alter tidal marsh habitat include diversion of freshwater inflow, a progressive rise in sea level, and contamination by toxic agents. Fortunately, various agencies are now working to reverse the trend of tidal marsh loss with ongoing wetland restoration efforts.

Outside of San Francisco Bay, tidal marsh is rare indeed. Unlike along the east coast or the Gulf coast, with their extensive coastal plains and fringing marshes, California's youthful geology, active tectonics, and relative aridity provide a precipitous coastline with few opportunities for tidal marsh development. Birds that depend on salt marsh habitats have few patches of habitat available.

Healthy tidal marsh provides a complex habitat for many fish and wildlife and is recognized as among the most productive habitats in the world. High primary productivity of tidal marshes results from contributions of nutrients from the tides as well as from the land. Marshes serve to filter and convert these nutrients and to build biological energy paths. It is a give-and-take relationship. As a result of vigorous plant growth, salt marshes produce copious amounts of *detritus*, the decomposition of which is a major pathway of energy back into the adjacent estuary. This "detritus pathway" may be more important than phytoplankton-based production as a basis for the estuarine food web. Amphipods, decomposers that abound in California's tidal marshes, are the primary food of many marsh-dwelling rails and sparrows, and the nutrients generated by tidal marshes feed the invertebrates that teem beneath the surface of adjacent tidal flats.

California Clapper Rail *(Rallus longirostris obsoletus)* and the California Black Rail *(Laterallus jamaicensis coturniculus)* populations are restricted almost entirely to San Francisco Bay's tidal

Plate 46. A Clapper Rail finds refuge on flotsam after being driven from its tidal marsh habitat by high spring tides. This periodic flooding increases the vulnerability of many marsh birds to predation.

marshes, as are three distinct geographic races of endemic salt marsh Song Sparrows *(Melospiza melodia)*—San Pablo *(M. m. samuelis)*, Suisun *(M. m. maxillaris)*, and Alameda *(M. m. pusillula)* Song Sparrows. The Song Sparrows illustrate evolution in action, and deserve a brief discussion here.

Among the most sedentary of California's birds, the tidal marsh Song Sparrows maintain territories year-round, although they actively defend them only in the breeding season, which, as with most resident birds, begins early in the season. However, these marsh-dwelling sparrows nest even earlier than the sparrows of adjacent upland habitats. Territorial birds are in full song by mid-February, and eggs are being incubated by mid-March. Because the marsh sparrows build their nests mostly in tidal marsh vegetation, often within a few inches of the ground, early nesting enables the sparrows to fledge young before spring high tides inundate the habitat each May and June. The other marsh nesters may also phase their nesting efforts to miss the high spring tides. Although their habitat is extremely restricted in extent, the marsh sparrows occur in extremely high densities (an average of about nine birds per acre!). The needs of a sparrow

Plate 47. The elusive Black Rail lives in the compact tangles of dense marsh vegetation. Rarely seen, its presence is often revealed only by its distinctive call.

Plate 48. Black Rails build well-hidden nests at ground level. Often nests are submerged by high spring tides, presenting an ever-present risk to the bird's breeding success.

pair and their offspring (they may produce two or three clutches in a season) are provided year-round within a fraction of an acre, an indication of the richness of the tidal marsh habitat.

The loss of coastal tidal marsh has been most pronounced in southern California, where only remnant patches remain. Morro Bay was formerly one of the few locations outside of the Bay Area that supported California Clapper Rail, but it has not been reported there since the mid-1970s. Black Rails still persist there in small numbers, however. Farther southward, below Point Conception, Light-footed Clapper Rail *(Rallus longirostris levipes)* replaces the *obsoletus* race, and, although gone from most of its former haunts, it can still be found in appropriate habitat in upper Newport Bay, the Tijuana River mouth, and patches of habitat in San Elijo Lagoon and south San Diego Bay. Another salt marsh *obligate,* the endangered Belding's Savannah Sparrow *(Passerculus sandwichensis beldingi),* still holds on as a permanent resident of the fragmented marshes from Santa Barbara southward to San Diego. Another distinctive race, the Large-billed Savannah Sparrow *(P. s. rostratus)* has a wintering range in California restricted entirely to tidal marshes from San Diego northward to Morro Bay.

Salt Pans and Ponds

Before they were so drastically altered by urbanization, California tidal marshes contained shallow "salt pans," ponds that were scattered along the backshore of the saline marshes. In San Francisco Bay, there were an estimated 8,000 acres of pans prior to 1850; today there are about 250 acres (Goals Project 1999). These pans were used by wading birds then as they are today; long-legged shorebirds such as avocets, stilts, and yellowlegs are particularly attracted to tidal marsh pans. Salt ponds are habitats created by people through the construction of levees on former tidal marshes and managed for salt production by commercial enterprises. Like salt pans, these ponds provide shallow, ponded wetlands that are used opportunistically by waterbirds. San Francisco Bay alone has more than 30,000 acres of salt ponds, the most extensive and heavily utilized salt pond complex for waterbirds in the United States. Ironically, this productive waterbird habitat overlies former tidal marsh (see above), and accounts for the displacement and endangerment of some of the native flora and fauna of the Bay Area. The creation of salt ponds, which followed the settlement of California in the mid-1800s, excluded some birds and attracted others. In the early 1900s, American Avocet and Black-necked Stilt were described as "typical inland species" whose breeding distribution reached the coast only southward of Santa Barbara. Avocets and stilts are now common breeders in the Bay Area, largely within the extensive shallow water habitat provided by these human-made salt ponds. In the early twentieth century, Western Sandpiper was described by Grinnell et al. (1918) as a "sparing winter visitant from San Francisco southward, but more numerous in southern part of the state." Now, Western Sandpiper is the most abundant shorebird wintering in the Bay Area; indeed, the bay holds an estimated two-thirds of the population wintering on the Pacific Coast. The development of these salt evaporation ponds, beginning around 1860 and proceeding through the 1930s, happened to correspond to the draining of some productive interior wetlands (e.g., Tulare Lake) and so provided alternative habitat as shorebirds were excluded from their interior foraging grounds. Conversion of tidal marsh to salt ponds accelerated from the 1930s through the 1950s. Terns and gulls have responded by establishing breeding colonies in the newly created habitat. Caspian Tern *(Sterna caspia)* began nesting in San Francisco Bay as early as 1922 and persists to the present. Forster's Tern, California Least Tern *(Sterna antillarum browni)*, and more recently Black Skimmer *(Rynchops niger)* followed suit, each species extending its coastal range northward by colonizing the levees and islands provided by the salt ponds. California Gull, which nests primarily east of the Sierras and the Cascades, established its first coastal breeding colony in 1982 in San Francisco Bay, also in response to the habitat created by the salt ponds and perhaps because of compromised habitat at Mono Lake. Nesting in San Francisco Bay coincided with a general range expansion in the Great Basin.

Other coastal breeding sites associated with salt ponds and of particular importance to terns (and Snowy Plovers) are Elkhorn Slough, Bolsa Chica (in round numbers: 4,000 pairs of Elegants, 300 Caspians, 220 Forsters, 50 Skimmers), Newport Back Bay, and south San Diego Bay (Western Salt Works). The latter site is notable for hosting a small colony of Gull-billed Tern *(Sterna nilotica vanrossemi)*, the only coastal population in the state (but see "The Deserts' Birds").

Although used in greatest numbers by shorebirds, gulls, and

terns, other waterbirds use salt ponds as well, even though their value as breeding sites is limited by salinity levels. Eared Grebe *(Podiceps nigricollis)* has bred in San Francisco Bay, and Eared Grebe along with American White Pelican *(Pelecanus erythrorhynchos)*, Double-crested Cormorant *(Phalacrocorax auritus)*, Northern Shoveler, Bufflehead, Canvasback, and phalaropes frequent ponds when water levels are deep enough and salinity levels low enough to attract or support prey.

Coastal Freshwater Marsh

Marshes in general and freshwater marshes in particular provide valuable bird habitat because standing water and saturated soil promote a biologically rich environment. Usually lying at the bottom of a watershed, marshes tend to accumulate high levels of nutrients and organic material that encourages prolific plant growth. The plants—cattails, tules, willows—provide dense cover for nesting and roosting and produce prodigious seeds. Insect and invertebrate life also thrives in the wetland environment,

Plate 49. In the spring, a Common Yellowthroat male sings to proclaim his chosen breeding territory. Yellowthroats belong to the wood-warbler family and should not be confused with another species from that family, the Yellow-throated Warbler.

Plate 50. Male Red-winged Blackbirds are polygynous. To attract females and define their territory, they perform a flamboyant wing-spreading display revealing their bright shoulder patches or "epaulets."

and the combined ingredients of cover and food support abundant birdlife.

California's relatively arid climate and the tectonic forces that fold and pleat much of the coastal slope provide limited opportunities for freshwater marshes to develop. Those marshes that did exist have also been depleted by the draining and filling of wetlands for agricultural, residential, and urban developments—especially around the periphery of the larger estuaries at San Francisco, San Diego, and Newport Bays. The freshwater marshes that remain are often degraded by grading, plowing, or grazing up to the wetland edge, thus reducing plant cover that is essential to birdlife. Fortunately, since the burgeoning environmental movement that began in the 1970s, some of these marshes, so valuable to California's native avifauna, have been and are being protected and restored. Those freshwater marshes that occur along the immediate coast attract a somewhat different complement of species than the marshes in the interior, and the rarity of this habitat increases its value to each species that relies on it for survival—Virginia Rail *(Rallus limicola)*, Sora *(Porzana carolina)*, Common Yellowthroat *(Geothlypis trichas)*, Marsh Wren *(Cistothorus palustris)*, Red-winged Blackbird *(Agelaius phoeniceus)*, and Tricolored Blackbird *(A. tricolor)*, among others. The rarity of freshwater marsh obligates—American Bittern *(Botaurus lentiginosus),* Least Bittern *(Ixobrychus exilis),* Cinnamon Teal, and Common Moorhen *(Gallinula chloropus)*—as breeding species along the immediate coast attests to the rarity of the habitat and, as Garrett and Dunn point out in *Birds of Southern California* (1981), can be "attributed to the large scale destruction of freshwater marshes."

Birds of Specific Coastal Wetlands

The following list of sites is by no means exhaustive, but merely representative.

North Coast

The Lake Earl–Lake Talawa wetland complex forms the largest coastal lagoon system in California and provides an important nesting and wintering area for a variety of waterfowl and marsh birds. This is one of the rare spots along the immediate north coast where American Bittern regularly nests. In fall, the open water provides an important coastal staging area for Canvasback. The lagoons are complemented by riparian forests as well as by islands of coniferous forest, and seasonally flooded pastures at the Smith River bottoms provide crucial wintering habitat for Aleutian Canada Geese *(Branta canadensis leucopareia).* Nearly the entire world population of this diminutive, white-collared, and formerly endangered race of this familiar species stages here in spring before migrating northward to the breeding grounds on the Arctic coast. Another at-risk population, the Western Snowy Plover *(Charadrius alexandrinus nivosus),* finds refuge on the sandy beach adjacent to Lake Talawa, where it occurs in relatively large numbers in winter.

Humboldt Bay also benefits from the proximity of diverse adjacent habitats—tidal marsh, the Mad River and Eel River estuaries, restored freshwater marsh (Arcata Marsh), coastal dunes (Lanphere-Christensen Dune Preserve), and extensive willow thickets and wet pasturelands. The eelgrass beds of Humboldt Bay attract the state's largest flocks of Brant in spring, when birds migrating from farther south stop to refuel en route to their Arctic breeding grounds. California's largest concentration of nesting and roosting colonies of Black-crowned Night-Heron on Indian and Teal Islands attest to the abundance of crabs and mollusks, the primary food of night-herons.

The Garcia River mouth coastal floodplain deserves mention because it attracts large concentrations of waterfowl in winter and spring. Most notably a flock of several hundred Tundra Swans arrive from breeding grounds in the high Arctic to winter here annually, one of only three coastal wintering grounds in the state. Flocks of several hundred of these elegant, long-necked, white giants can be seen from Hwy. 1 as they graze in the plowed fields. Some immatures are washed gray, but by spring the entire flock is almost immaculate, save for the face, which is sometimes stained rusty from grubbing for tubers and roots in the soil.

Central Coast

San Francisco Bay, containing 90 percent of the coastal wetlands in California, encompasses a vast network of tidal and diked baylands, tidal flats, emergent tidal marsh, salt ponds, diked wetlands, agricultural bayland, and moist grassland. Associated wetland habitats include vernal pools, riparian forest, and channels, creeks, sloughs, and canals. All totaled, it includes over one-half million acres of wetland habitat (see Table 2).

Subregions within the bay—Suisun, North (San Pablo) Bay, Central Bay, and South Bay—each has distinctive habitat values and supports its own unique complement of birdlife. San Francisco Bay supports the largest population of Canvasbacks on the Pacific Flyway and is the most important wintering area in North America. The open-water habitats also support large concentrations of other diving waterfowl, most commonly, in order of abundance, Greater Scaup, Surf Scoter, Ruddy Duck, and Bufflehead. The wintering population of Ruddy Duck is one of the largest in North America, and the Bay Area is the most important wintering habitat for Surf Scoter south of Puget Sound. Dabbling ducks are less abundant than diving ducks in the bay because of the relative proximity of shallower water habitats upstream in the Sacramento–San Joaquin Delta and the Central Valley wildlife refuges; however, the managed wetlands of Suisun Bay, in particular, offer breeding habitat for Mallard, Gadwall, and Northern Pintail.

TABLE 2. Present Habitat Acreage for San Francisco Bay Estuary

Habitat Type	Historic	Modern
Open water: bay and channel	273,971	254,228
Tidal flat	50,469	29,212
Tidal marsh	189,931	40,191
Lagoon	84	3,620
Salt pond	1,594	34,455
Diked wetland	0	64,518
Agricultural bayland	0	34,620
Storage or treatment ponds	0	3,671
Undeveloped bay fill	12	7,598
Developed bay fill	0	42,563
Other baylands	254	1,951
Total acreage	**516,315**	**516,625**

After the Goals Project 1999.

The salt ponds in the South Bay and structures such as the abandoned airfield at Alameda National Wildlife Refuge in Oakland provide advantageous habitat in areas that were created for other purposes. Since being abandoned by the Navy, the Alameda airfield has been colonized by Caspian Terns and California Least Terns as nesting habitat and has supported important colonies, at least in some years.

Elkhorn Slough in Monterey County and Morro Bay in San Luis Obispo County are two exceptionally important coastal wetlands south of the San Francisco Bay, especially for migrating shorebirds. Elkhorn Slough, about halfway between Santa Cruz and Monterey, is a 1,000-acre inlet of sandy tidal flats, salt marsh, and sloughs surrounded by abandoned salt evaporation ponds and flooded bottomlands. Dabbling ducks—Mallard, Gadwall, Northern Pintail, and Northern Shoveler—nest in the marshlands, and Snowy Plover and Forster's and Caspian Terns nest in small numbers on the salt flats. Morro Bay, which provides a large, protected, lagoonlike area of open water as well as tidal flats and salt marsh, is complemented by the oak and riparian woodland within Morro Bay State Park and the Morro Dunes Natural Preserve.

Both Morro Bay and Elkhorn Slough supported outlier populations of California Clapper Rail up until the 1970s, but this endangered species has not been seen at either site in recent years. Morro Bay, however, still does support a small population of California Black Rail, one of only a few outer coastal wetlands that do; others include Bolinas Lagoon, Tomales Bay, and Bodega Bay. Given their small size, limited habitat, and relative isolation, these satellite populations are vulnerable to extirpation.

South Coast

Mugu Lagoon, at the mouth of Calleguas Creek, Ventura County, is one of the most important migratory stops on the southern coast for shorebirds, especially during spring, when upward of 60,000 have been counted using the 2,000-acre wetland. Both shorebirds and waterfowl occur in good numbers in winter, and it is one of the most important roosting sites for Brown Pelican in southern California. The 200-acre salt marsh at Ormand Beach, just north of the lagoon, is a remnant of a once more extensive habitat, but it still supports Light-footed Clapper Rail and is a population center for Belding's Savannah Sparrow, a south coast salt marsh specialist. Adjacent agricultural pastures at Oxnard Plain provide additional secondary habitat for shorebirds when tidal flats are flooded, most notably wintering flocks of Pacific Golden-Plover *(Pluvialis fulva)*, a rarity in California.

Orange County is heavily urbanized, but several estuaries still retain their natural hydrodynamic functions and so serve as important bird habitats. These include Bolsa Chica and upper Newport Bay, where islands support large colonies of nesting terns, most notably the endangered California Least Tern and Black Skimmers. A major colony of Elegant Terns (4,000 pairs) has established itself at Bolsa Chica since the late 1970s. Seal Beach National Wildlife Refuge in Anaheim Bay is another site that attracts nesting terns and migrating shorebirds.

Mission Bay and San Diego Bay, San Diego County, are urbanized estuaries, highly impacted by human activity yet still providing habitat for many of the same birds found at other southern California wetlands. Brant congregate in winter and spring, foraging on eelgrass beds, and California Least Tern began nesting on sandbars especially constructed for the purpose in the 1970s.

BIRDS OF THE COAST RANGES

The three fieldnote excerpts (see sidebar "Journal Excerpts") are all from the California Coast Ranges, as complex a bioregion as can be found perhaps anywhere in the world. Given California's size (158,706 square miles, spanning 10 degrees of latitude) and its powerful oceanic influences (see "Seabirds and the Marine Environment"), there is extravagant variation in climate from north to south. On the coastal slope, the northern third of the state is relatively humid, with high precipitation that can occur year-round, whereas the central and southern coast are regions of winter rain and summer drought. Precipitation in the redwood region on the northernmost coast averages 70 inches annually; farther inland, in the wettest parts of the Siskiyous, it increases to 140 inches. Rainfall generally decreases southward, as representative annual rainfall averages for coastal locations in each county illustrate: Mendocino, 35 inches; Monterey, 18 inches; and San Diego, 10 inches. There are, of course, exceptions to this general trend; areas of Big Sur on the Monterey coast are nearly as rainy as the Siskiyous. The northern third of the state experiences winter frost regularly, and freshwater ponds may freeze, though

Map 5. Coast Ranges Bioregion.

Plate 51. Hutton's Vireos are among the earliest of spring nesters. They breed in the dappled oak–bay woodlands of the coastal ranges.

the ground does not. South of Mendocino, frost is uncommon at lower elevations, especially near the immediate coast.

Because of the complex topography of the Coast Ranges, there is also a large differential in rainfall from the coast as you move upslope from the coastal plain or inland across the mountain ridges. This variation is reflected, often dramatically, in the plant communities. Because of the marine influence—the relatively moist air that the ocean provides and its moderating effect on temperature—rainfall distinguishes the seasons more sharply than does temperature, especially in the southern two-thirds of the state, where almost all the rain falls from November through March. As you move inland from the outer to the inner Coast Ranges, rainfall decreases significantly. This general trend is sometimes contradicted by the local topography on the immediate coast. For example, on the Point Reyes Peninsula, the outermost coastal plain receives only about 19 inches of precipitation a year. As air currents rise over the first coastal ridge, cooling and condensation wrings out the moisture, and so within a few miles, on the east slope of the first ridge, annual rainfall more than doubles.

The Coast Ranges are essentially terrestrial habitat, so their birdlife is characterized more by *passerines* (landbirds) and soar-

(continued on page 120)

JOURNAL EXCERPTS

MID-MAY, SUNRISE. NORTH COAST. PRAIRIE CREEK REDWOODS.

Overhead we hear the distinctive "kleeer-kleeer-kleeer" calls of Marbled Murrelets as they commute between nesting sites in the old-growth redwood grove and feeding grounds in the open ocean. The coastal fog hangs low over the meadow—a herd of Roosevelt Elk feed silently, heads bowed in the soft morning light. Soon we hear the strange, mechanical "veeeer" of a Varied Thrush from the deep redwood-spruce forest that frames the open pasture, then a Steller's Jay's insistent scolding (of what? Spotted Owl? Barred Owl?). As the fog lifts and the sun breaks through, the clarion song of a White-crowned Sparrow and the chatters of Chestnut-backed Chickadees and Winter Wrens greet the day.

Plate 52. The presence of a Canyon Wren in its rocky cliff habitat is often announced by its evocative call, a cascading series of melodious, liquid notes.

LATE OCTOBER, LATE AFTERNOON. INNER COAST RANGES. PINNACLES NATIONAL MONUMENT.

As we walk down the eastern slope, a fence lizard dashes across the trail, hellbent, a Greater Roadrunner right behind. A Canyon Wren trills from an outcropping; juncos flush-up, flashing white tail feathers. A California Thrasher whistles in the distance, followed by the familiar "ping-pong" song of a Wrentit. Acorn Woodpeckers work their oak plot, and we hear the querulous call of a Hutton's Vireo. As the landscape loses definition in the fading light, the plaintive whistle of a Northern Pygmy-Owl punctuates the dusk.

Plate 53. The distinctive call of the Wrentit is often the only clue to the presence of this secretive denizen of tangled scrub. It is a year-round resident in chaparral and coastal scrub and endemic to western coastal states.

Plate 54. A Wrentit, released after being banded, exhibits feather wear imposed by the rigors of breeding activity.

EARLY SPRING, MIDDAY. SOUTH COAST. TEMECULA CREEK, SAN DIEGO COUNTY.

In the drier reaches of the southern Coast Ranges, the desert environment approaches the shoreline, mixing with hard chaparral. Rufous-crowned Sparrow, a sage scrub species, nests in the same neighborhood as Black-throated Sparrow, a bird of the desert scrub. Other emblematic desert species such as Scott's Oriole, Black-tailed Gnatcatcher, and Gambel's Quail have found their way coastward here and consort with their coastal cousins—Bullock's Oriole, California Gnatcatcher, and California Quail.

Plate 55. In early spring, a Western Scrub-Jay collects rootlets to line its cup-shaped nest.

ing raptors than by the waterbirds that dominated the previous two parts of this book.

Because of the moderating influence of the marine air, spring is the longest season in the Coast Ranges. Along the central coast, spring begins as early as December or January, when resident birds such as Anna's Hummingbird *(Calypte anna)* begin their breeding display and Western Scrub-Jay *(Aphelocoma californica)* may already be nesting. By late January, Migratory Allen's Hummingbirds *(Selasphorus sasin sasin)* have returned from

Plate 56. High-speed photography reveals the elegance of a Warbling Vireo in flight.

Plate 57. The Pacific-slope Flycatcher usually builds its moss-covered nest in shrubbery beneath the forest canopy, but may also choose locations such as in the shelter of an undercut bank or the eaves of a building.

Plate 58. The Swainson's Thrush is a summer visitor to coastal woodlands, which often resonate with their ascending flutelike songs, among the most beautiful to be heard in California.

Central America and are foraging among flowering manzanita. By February, Hutton's Vireo *(Vireo huttoni)*, California Towhee *(Pipilo crissalis)*, and Purple Finch *(Carpodacus purpureus)* are singing emphatically. In March, *neotropical migrants* such as Warbling Vireo *(Vireo gilvus)*, Wilson's Warbler *(Wilsonia pusilla)*, and Pacific-slope Flycatcher *(Empidonax difficilis)* return to their breeding territories. By mid to late April, even the latest migrants—Swainson's Thrush *(Catharus ustulatus)* and Yellow Warbler *(Dendroica petechia)*—have returned. Landbird migration continues well into June, with a surge of northerly breeding insectivores pushing through during the last week of May, and

Plate 59. The songs of Song Sparrows vary widely among individuals and populations, but all variations are characteristic of California's wetlands.

even into June, when the latest breeding species, such as California Quail *(Callipepla californica)* and American Goldfinch *(Carduelis tristis)*, are finally fledging chicks.

Fall, too, is protracted throughout the Coast Ranges. Migratory movements of shorebirds ("Birds of the Shoreline"), raptors, and landbirds begin in mid-July and continue well into October with a peak in mid- to late September. The Coast Ranges' seasons, like the habitats, overlap intricately; spring overlies winter and morphs into fall, barely stopping to acknowledge summer, which, especially to the north, is often hidden beneath a layer of coastal fog. Winter can be assertive, but rainy periods may be interspersed with bright, warm days accompanied by the singing of Purple Finches, Winter Wrens *(Troglodytes troglodytes)*, and Song Sparrows *(Melospiza melodia)*.

Subregions

The Coast Ranges are within the California Floristic Province as defined in the *Jepson Manual* (Hickman 1993), which encompasses a half dozen subregions directly influenced by the coastal climate. Collectively, these coastal areas are called the *cismontane*.

Birds, being more mobile than plants and more adaptable to a wide range of environmental conditions, are not strictly confined to floristic communities, and most birds respond to fairly broad environmental variables.

Many species, however, do have distinct affinities for certain plant communities and their structure. For example, Acorn Woodpecker *(Melanerpes formicivorus)* may use a variety of species of oak trees, but tends to be associated with relatively mature and open oak woodlands. Steller's Jay *(Cyanocitta stelleri)* may occur in dense conifer forest or mixed evergreen forest, but responds to the density of the forest rather than to the tree species present. California Towhee occurs in a wide variety of shrub habitats, but needs some open space between the bushes and a fair proportion of bare or sparsely vegetated ground.

For the purposes of this book, we use the term "humid north coast" to refer to the stretch from the Oregon border southward through Mendocino County. The "central coast" refers to the nine counties that extend to the shores of San Francisco Bay as well as Santa Cruz, Monterey, and San Luis Obispo Counties, and the "arid south coast" extends from Santa Barbara County southward to the Mexican border.

The Coast Ranges and foothills of California are a topographically and climatically diverse region, and include conifer forests, mixed evergreen forests, oak woodland and savannah, coastal scrub, arid chaparral, grasslands, prairie, and riparian corridors. This potpourri of coastal habitats known as the cismontane encompasses one of the earth's hot spots of biodiversity. Well over half the landbirds occurring in California frequent some Coast

Plate 60. A male California Quail, acting as lookout, can often be seen perched above a group of quail foraging in the undergrowth.

Map 6. Geographic subregions of the Coast Ranges.

Ranges habitat; therefore, only the most characteristic species are discussed here.

North Coast

The humid north coast is dominated by conifer forest or mixed evergreen forest with some grassland and prairie interspersed. To the north (Del Norte and northern Humboldt Counties) the Coast Ranges extend inland to the Klamath Mountains eastward, and then merge with the Cascades. The Klamath Mountains share affinities with the Sierras and Cascades, but they are also influenced directly by the coastal climate regime. The mixed alliances of the Klamath province are illustrated by the distributions of two closely related species *(congeners):* Chestnut-backed Chickadee *(Poecile rufescens)* and Mountain Chickadee *(P. gambeli)*. Chestnut-backs are residents in the coastally influenced environments of the North and Central Coast Ranges; to the north their range extends inland and eastward into the Klamath and

Plate 61. The Chestnut-backed Chickadee inhabits humid coastal forest. Its nest is usually in a tree cavity accessed through an inconspicuous entrance, or it may use a convenient nest box.

Siskiyou Mountains. Mountain Chickadees, in contrast, are absent from the coast (except for an outlier population in the Santa Lucia Mountains) and range throughout the Sierras and Cascades; in the north, however, they extend westward into the inner Coast Ranges of the Klamath Mountains, where they overlap with Chestnut-backs.

When the ranges of such similar species overlap, it suggests that each is attuned to different niches within the habitats of the region. The distributional differences of these closely related chickadees mirror their individual evolutionary histories. Although we do not fully understand those differing stories, it is likely that the two species came from a common ancestor that diverged into separate species when climatic conditions (the Pleistocene Ice Age—1.6 million to 10,000 years ago) isolated each from the other—one toward the low-elevation coastal forests and the other into the high-elevation mountain forests. Over time, these isolated populations adapted to these disparate environments and developed plumage, morphology, and behavior that was beneficial to surviving in their respective habitats—the Chestnut-backed in the humid coast zone, the Mountain Chickadee in the montane zone. When the climate changed, allowing the two relatives to come back into contact, they had developed into separate species. This pattern of diversification of a group of relatives *(taxon)* into several ecological roles or modes of life is known as "adaptive radiation," an expression of evolution's creativity and genius. The chickadees (several species in the genus *Poecile*) provide a vivid example of what is called the "taxon cycle," in which a species spreads while adapted to one environment and then becomes restricted in its range while adapting to another environment. A third species in the genus *Poecile,* the Black-capped Chickadee *(Poecile atricapilla),* is very local in the northwestern counties, where it is associated mostly with riparian habitats.

Coastal Del Norte and Humboldt Counties are forested with mature stands of conifers, especially coast redwood *(Sequoia sempervirens),* Sitka spruce *(Picea sitchensis),* and Douglas-fir *(Pseudotsuga menziesii).* These moist forests attract birds found most commonly along the northwest coast—Barred Owl *(Strix varia),* Ruffed Grouse *(Bonasa umbellus),* and Gray Jay *(Perisoreus canadensis).* North coast forests do not end abruptly at southern Humboldt County, however, but continue intermittently as microclimate allows, southward to Santa Cruz and even into Monterey County. This narrow coastal forest, with enough winter rainfall and summer fog to support remnant groves of redwoods, conforms closely with the breeding distribution of Vaux's Swift *(Chaetura vauxi)* and Northern Spotted Owl *(Strix occidentalis caurina)* in California.

Northern Spotted Owl, characteristic of the dense, fog-drenched forests, is rarely noticed because of its strictly nocturnal habits and its reserved demeanor. Because of loss of habitat resulting from aggressive forestry practices in older growth forests, it is a threatened species in California. It faces a secondary threat as well. Studies in Oregon and Washington have found that where the forest has been fragmented by logging, the Barred Owl has encroached into territory formerly held by the Spotted Owl. Although Spotted Owls require dense forest, Barred Owls are more tolerant of open forests and thus are more adapted to the environment being created by forestry practices. Barred Owls were first detected in the northernmost Coast Ranges of California in the early 1980s. By 2002, they had been found as far south as Muir Woods National Monument in Marin County, the southernmost extent of the Northern Spotted Owl in the humid north

Plate 62. During daylight hours, the Spotted Owl may be found roosting in dense forest.

coast forest. The theory of competitive exclusion suggests that two such similar species cannot coexist, that ultimately one will displace the other. Larger and adapted to more open habitat than the Spotted Owl, Barred Owl seems to have the competitive advantage, given California's current forestry practices.

Marbled Murrelet *(Brachyramphus marmoratus)* is a small seabird with the curious and unusual habit of nesting high in the limbs of old-growth conifer forests. Its nesting habits are so unusual for a seabird that they remained a mystery until 1974, when a nest site was discovered in Big Basin State Park, Santa Cruz County. Except for the apparently small outlier population in the Santa Cruz Mountains, the entire breeding range of the Marbled Murrelet in California is encompassed by the conifer forests on the immediate coast of Del Norte, Humboldt, and northern Mendocino Counties, a habitat that is threatened by the same practices that threaten the deep-forest-dwelling owls and other north coast species.

North coast riparian habitat provides important habitat for neotropical migrants such as Pacific-slope Flycatcher, Warbling

Vireo, Wilson's and Yellow Warblers, and Swainson's Thrush, both for breeding and migration. Patches of riparian forest on the immediate north coast along the Mad and Eel Rivers, also holds breeding Black-capped Chickadee and American Redstart *(Setophaga ruticilla),* two species absent (except very sporadically) elsewhere on the coast. The chickadee is a montane breeder with a distribution extending into California through the Klamath Mountains and coastward only as far south as Humboldt County. The Redstart is an anomaly, as this is one of few breeding sites in the state (others have been found at Big Sur and Point Reyes). Among its family, the wood-warblers *(Parulidae),* this species is highly migratory and occurs regularly, though rarely, with some reliability at coastal migrant traps. Redstart is one of the three most common vagrants at the Farallon Islands, along with Blackpoll Warbler *(Dendroica striata)* and Palm Warbler *(D. palmarum).* These pioneer vagrants, usually juvenile birds, probably suffer very high mortality during their first off-course migration. However, on occasion, a wandering individual or two may find appropriate habitat outside the species' normal range and stay to breed, establishing a new outlying breeding colony. The resident Redstarts of Humboldt likely originated with such pioneers.

Central Coast

The central coast includes the Bay Area and extends eastward through Suisun Bay to the Sacramento–San Joaquin Delta. The lowlands of the Delta, however, are considered allied with the Central Valley. With few exceptions, the Central Coast Ranges are relatively low in elevation (less than 4,000 feet) and trend parallel to the coast in a series of folded ridges. In general, rainfall decreases as marine air moves from west to east, so the outer Coast Ranges are much more moist than the inner Coast Ranges. Marine terraces are fairly common on the immediate coast, backed by a coastal slope that begins to rise slightly inland. Higher peaks in the Coast Ranges (e.g., Junipero Serra in Monterey County at 5,862 feet) introduce a limited montane zone within the Coast Ranges; however, the central coast has less of this montane habitat than the north or south coasts. The central coast is a zone of overlap between northern and southern influences, highly susceptible to variations in microclimate (caused by slope effect and

Plate 63. California Thrashers, usually secretive by nature, will occasionally maintain a lengthy monologue of chatter from a prominent perch.

aspect, *edaphic* conditions, etc.). Here, coast redwoods may grow in a canyon bottom that harbors Northern Spotted Owls, whereas the adjacent south-facing hillside may be shrubby chaparral, hosting Wrentits *(Chamaea fasciata)* or California Thrashers *(Toxostoma redivivum).*

The immediate coast, from Point Reyes southward to Santa Barbara, almost entirely frost free in winter and generally dry from April through October, is one of the most temperate and equitable climates in the world and hosts as diverse an avian community as anywhere in North America. The coastal counties of Marin, Monterey, and Santa Barbara, with their complement of large areas of protected open space, coastal mountains, grasslands, riparian thickets, marshes, conifer and mixed evergreen forests, and well-tended ornamental gardens, have county bird lists rivaling or surpassing those of most states: Marin, 481 species; Monterey 484; Santa Barbara 485. And each is gaining species yearly.

Nuttall's White-crowned Sparrow *(Zonotrichia leucophrys nuttallii)* is a year-round resident along the immediate coast; in spring its sweet song rings from the lupine-covered dunes and coastal scrub within the fog belt. Other more northerly races, Gambel's and Puget Sound White-crowned Sparrows *(Z. l. gam-*

Plate 64. A subspecies of the White-crowned Sparrow, *Zonotrichia leucophrys nuttalli,* is common year-round in scrub within the fog belt of the Coast Ranges.

belii and *Z. l. pugetensis),* arrive to spend the winter, but Nuttall's stays year-round, a testament to the equitability of the climate. The song of the White-crowned, a signature sound of the coastline, is very "site specific." The late Luis Baptista spent a full life studying this beautiful phrasing and discovered that each *deme* (a locally distinct population) had its own dialect that was recognizable, not only to Baptista's well-tempered ear but also to other members of the deme. So, if you were to walk along the coast from Bodega Bay to Bolinas, say, in March, every few miles you would hear a distinct variation on the basic song of the resident White-crowned Sparrows. Usually the last few notes of the song vary, but in a highly recognizable way. These dialects do not change gradually, but rather abruptly; so the Bodega dialect sounds nothing like the Bolinas dialect, which in turn is distinct from the Golden Gate Park or the Berkeley song. These sparrows are nonmigratory (sedentary) and therefore evolutionary products of their coastal neighborhoods; through their songs, they express the subtle variety that exists in their natural habitat, in this case the soft chaparral of the central coast.

In this book, by categorizing habitats and species, we are painting with a broad brush. By paying attention to the nuances of sparrow song, we come to understand that nature cannot be easily parsed, that a subtle variety underlies our distinctions, and that when left to its own devices, its own exuberance, the natural world is infinitely creative.

Raptors—hawks, falcons, kites, and eagles—migrating southward along the north-south trending Coast Ranges of California follow a natural corridor of rising currents and thermals in their passage. These rising air currents assist the effort of flight and reduce the energy requirements of migration. Autumn hawk migration is most dramatic along the central coast at Marin Headlands, located directly upslope from the Golden Gate Bridge. Hawks are hesitant to cross open water because air currents and thermals are dampened and humid air is relatively heavy. The outer Coast Ranges converge just north of San Francisco Bay, in effect funneling birds that are following the ridges onto the flanks of Mount Tamalpais, just north of the Golden Gate. Each fall, from August through mid-November, birders congregate at the Marin Headlands (Point Diablo) to watch and count the river of raptors overhead. Lull periods may be followed by a flurry of activity, depending on the conditions. Peak passage tends to occur midday (10:00 a.m. to 2:00 p.m.) on relatively clear, relatively calm days when the coastal fog is not too thick, especially during September.

The phenomenon was first noticed by Laurie Binford during the 1970s, when he was curator of ornithology at the California Academy of Sciences in Golden Gate Park. From his office window during fall, Binford noticed an inordinate number of hawks, especially *accipiters*, flying by, heading southward. Knowing these were migrants coming from the north, he conjectured that they were crossing San Francisco Bay at its narrowest point. Several well-timed field trips to the top of the Marin Headlands, which afforded vast views of the surrounding ridges and the Golden Gate Bridge, confirmed Binford's suspicions. This discovery generated much enthusiasm among the local birding community, not only because the common raptors—Red-tailed, Cooper's, and Sharp-shinned Hawks *(Buteo jamaicensis, Accipiter cooperii,* and *A. striatus)*—were expected, but because other rarer species were noted during Binford's initial reconnaissance trips. Soon, the observations were organized systematically, with observers present during most daylight hours. This led to the founding of the Golden Gate Raptor Observatory in 1992, which,

TABLE 3. Raptor Sightings at Marin Headlands

Species	Hawks per Hour*	Average Annual Number 1992–2002
Turkey Vulture	21.10	5,554
Osprey	0.20	61
White-tailed Kite	0.15	33
Bald Eagle	<0.01	0
Northern Harrier	2.52	523
Sharp-shinned Hawk	11.16	3,420
Cooper's Hawk	5.29	1,922
Northern Goshawk	<0.01	1
Red-shouldered Hawk	1.29	192
Broad-winged Hawk	0.34	81
Swainson's Hawk	0.02	2
Red-tailed Hawk	23.29	6,597
Ferruginous Hawk	0.06	15
Rough-legged Hawk	0.02	5
Golden Eagle	0.04	13
American Kestrel	1.28	482
Merlin	0.40	105
Peregrine Falcon	1.20	77
Prairie Falcon	0.02	4
Unidentified	2.59	1,575
Total	**70.99**	**20,662**

* Average annual hours of observation = 413.

From Golden Gate Raptor Observatory 2002.

with the help of a cadre of dedicated volunteers, has kept yearly records that provide valuable insights into hawk migration, relative abundance, annual variability of population size, timing, and health (see Table 3).

Relative abundances of these long-term observations accurately reflect the commonness and rarity of each species, except perhaps for Broad-winged Hawk *(Buteo platypterus)*, which is encountered relatively rarely elsewhere in the state.

South Coast

The southern coast extends from San Luis Obispo southward to the Mexican border and includes the west-to-east trending Transverse Ranges (San Gabriel, Santa Ynez, Santa Monica, and San Bernardino Mountains), the Peninsular Ranges (most notably the San Jacinto Mountains), and the interspersed southern coast lowlands. Most of the lowlands have been urbanized, and only remnant wetlands and grasslands remain. Most native habitat remaining is coastal (or "marine") chaparral made up of a Mexican, *Madro-Tertiary* flora dominated by drought-tolerant, fire-adapted shrubs with scrub oaks *(Quercus dumosa)*, cacti, and mesquite intermixed. This is not only an area of overlap between outer and interior California but also of overlap between Alta and Baja California. The coastal slopes along the southern coast tend to rise abruptly from the shore, and some peaks are quite high (more than 10,000 feet), adding montane habitats and bird species otherwise absent from much of the central coast

In Riverside and San Diego Counties, within the southernmost reaches of the Coast Ranges, desertlike environments approach the immediate coast and merge with chaparral scrub habitat. Where natural habitats still retain their essential character, places such as the southwestern slopes of the San Jacinto Mountains, the vegetative community is diverse—live-oak woodlands intermixed with coastal sage scrub, hard chaparral, cottonwood–willow riparian, and mesquite dry washes. Coastal birds nest in the same neighborhood as species usually associated with desert environments. This a zone of sympatry between two sibling species—California and Black-tailed Gnatcatchers *(Polioptila californica* and *P. melanura)*—that were, until recently, considered subspecies of the same species. The California Gnatcatcher is a bird of the coastal scrub; the Black-tailed is thoroughly desert adapted. Other emblematic desert species such as Black-throated Sparrow *(Amphispiza bilineata)*, Scott's Oriole *(Icterus parisorum)*, and Gambel's Quail *(Callipepla gambelii)* have found their way coastward here and consort with their coastal relatives, such as Bell's Sage Sparrow *(Ampispiza belli belli)*, Bullock's Oriole *(Icterus bullockii)*, and California Quail.

The Carrizo Plain, though technically within the southern coast subregion, shares characteristics with both the coast and the deserts but is most closely associated with the San Joaquin Valley, with which it is treated (see "Birds of the Central Valley and Delta").

Habitats

From the rain-forest environment of the Siskiyou Mountains, with rainfall averaging 140 inches per year, to the desert conditions on the coastal slopes of the Peninsular Ranges, with an average of 10 inches a year, the Coast Ranges comprise as varied a mosaic of habitat types as any California bioregion. Some of these habitats are as biologically endowed as any in the world; others are depauperate, supporting only a few hardy species. The following habitat categories are generalizations meant to inform the curious naturalist about what birds to expect where.

Coastal Cliffs and Promontories

Along much of the immediate coast, especially at headlands, the Coast Ranges have been sculpted by the combined forces of erosion and *plate tectonics* into dramatic escarpments and precipitous cliff faces. Few plants take hold here, but fissures and seeps, ledges and outcroppings afford bold opportunities for a handful of tenacious bird species. Famously, some of these ledges and caves have been used to *hack* Peregrine Falcons *(Falco peregrinus)* that were raised from the egg in the laboratory. The captive breeding program is part of an effort to rescue this top-level predator, which is having limited success breeding in the wild because of pesticide contamination. Previously, and now with some regularity, Peregrines are using these sites for nesting of their own accord—one of the more successful stories of active wildlife management. Other raptors—Great Horned Owl *(Bubo virginianus)* and Barn Owl *(Tyto alba)*—also occupy crevices and crannies in cliffs along the beach, both for nesting and roosting. Common Raven *(Corvus corax)*, a premier omnivorous scavenger, uses coastal ledges for nesting and as a vantage point, ever watchful for shorecast carrion—dead gulls, seals, fish—or perhaps a Snowy Plover *(Charadrius alexandrinus)* nest.

Except for the aptly named Rock Wren *(Salpinctes obsoletus)*,

Plate 65. The Great Horned Owl finds a suitable cave, cliff crevice, or an abandoned hawk nest in which to nest.

which favors fractures, fissures, and talus slopes, few passerines nest on the barren coastal cliffs. But occasionally, when a small freshet provides moist mud, Cliff Swallows *(Petrochelidon pyrrhonota)* construct their mud nest colonies beneath an overhanging rocky shelf. Bank Swallows *(Riparia riparia)*, too, though much rarer, nest at a few coastal escarpments, most notably on the Smith River, Del Norte County, and formerly at Fort Funston in San Francisco. Another aerial insectivore, the Black Swift *(Cypseloides niger)*, a relatively rare species in California usually associated with remote mountain waterfalls, is known to nest in sea bluffs above the surf at just a few locations between San Mateo and San Luis Obispo.

Coastal Prairie, Strand, and Dune Scrub

Buffeted by wind and spray, few plants develop on the shoreline fore dunes, but in the hind-dune shadow, communities of yellow bush lupine *(Lupinus arboreus)* and other shrubs may take hold. Few birds are hardy enough to stay here year-round, but White-crowned Sparrows *(Zonotrichia leucophrys)* seem to manage, and many passerines move through during migration. Gnatcatchers and goldfinches, and House Wrens *(Troglodytes aedon)* and Yellow-rumped Warblers *(Dendroica coronata)* are particularly fond of these lupine patches, through which they forage exploring the foliage for insects. Golden-crowned Sparrows *(Zonotrichia atricapilla)*, Fox Sparrows *(Passerella iliaca)*, Spotted Towhees *(Pipilo maculatus)*, and even Downy Woodpeckers *(Picoides pubescens)* visit, too, especially in fall.

Most native birds are adapted to native habitats; as the structure of the community changes, many highly refined species are excluded from territories they formerly occupied. Originally covered by perennial native bunchgrass, Douglas iris *(Iris douglasiana)*, and low-growing shrubs, intensive grazing and housing developments have converted much of the coastal prairie on marine terraces along the immediate coast to fields of weedy (nonnative, annual) grasses. In areas where coastal prairie is still intact, communities of native birds may still be found; where coastal prairie has been replaced by pasture and converted to annual grasslands, a less complex community of birds dominates. In early spring, when the Douglas iris are blooming, the electric, buzzy song of the Savannah Sparrow *(Passerculus sandwichensis)* and the robust flutelike song of the Western Meadowlark *(Sturnella neglecta)* announce these most common residents of the coastal prairie. Low, moist swales may attract nesting Northern Harriers *(Circus cyaneus)* and sometimes Common Yellowthroats *(Geothlypis trichas)*. When lupine or coyote brush *(Baccharis pilularis)* is present, California Quail and American Goldfinch occur, and when nearby trees or fence posts provide nesting or observation sites, American Kestrel *(Falco sparverius)*, Western Bluebird *(Sialia mexicana)*, or Western Kingbird *(Tyrannus verticalis)* may be present. Common Ravens patrol overhead, along with the Coast Ranges' ubiquitous Red-tailed Hawk and Turkey Vulture *(Cathartes aura)*. In the fall and winter, open-country-loving birds such as Say's Phoebe *(Sayornis saya)*, Loggerhead Shrike *(Lanius ludovicianus)*, Short-eared Owl *(Asio flammeus)*, and even Burrowing Owl *(Athene cunicularia)* may show up; however, all but the phoebe are becoming scarce almost everywhere in California. These population declines are indicative of the conversion and destruction of grasslands and damp meadow environments that is rampant throughout interior California. Coastal prairie, although attractive to these species in the nonbreeding season, is less used during breeding. Where annual grasses have replaced bunchgrasses, and where shrubs are absent, the bird community is diminished; however, the ground-nesting Savannah Sparrow and Western Meadowlark seem to be holding on.

Chaparral

Habitats dominated by shrubs and small trees, sometimes derisively called "scrub lands," are lumped generically under the term "chaparral," from the Spanish *chaparro* meaning evergreen or live oak, probably derived from the Basque word *txapar,* meaning thicket. (The special trousers worn by vaqueros for protection when riding through these thorny thickets were called "chaps," after the Mexican word of the same origin, *chaparreras.*) Chaparral is a drought- and fire-adapted plant community, often growing on thin and nutrient-poor soils. Various types of chaparral represent one of California's most common plant communities, covering an estimated 12 million acres, or about 12 percent of the state. Chaparral occurs in the Sierra foothills and the desert regions as well, but the preponderance of chaparral habitat is in the Coast Ranges, particularly in the more arid regions of central and southern California. In San Diego County, for example, chaparral covers about 35 percent of the landscape.

Chaparral can be split broadly into two types—"soft" and "hard" chaparral. Soft chaparral is generally at lower elevation along the immediate coast, where there is more moisture in the air and lower *evapotranspiration* rates. Coastal scrub and coastal sage communities are each called "soft chaparral" because the dominant shrubs—coyote brush to the north, California sage *(Artemisia californica)* to the south—are relatively soft to the touch. Soils tend to be alluvial with higher nutrient content than those of shrub associations at higher elevations. The breeding distribution of Allen's Hummingbird *(Selasphorus sasin)* and Nuttall's White-crowned Sparrow *(Zonotrichia leucophrys nuttalli)* conforms almost exactly to the distribution of soft chaparral.

Distinct communities of birds are associated with both soft and hard chaparral; those of the hard chaparral tend to be more strict in their life activity requirements and more limited in their distribution than those of the soft chaparral. For example, the migratory Allen's Hummingbird does quite well in suburban en-

Map 7. Breeding distribution of Allen's Hummingbird (after Grinnell and Miller 1944).

vironments and habitat edges within the coastal zone. Hard chaparral species are much more localized; California Thrasher, for example, rarely wanders beyond the confines of its arid habitat boundaries.

North coast chaparral, found intermittently from southern Oregon southward to San Francisco Bay, intergrades with northern coastal prairie as edaphic and climatic conditions vary. North coast chaparral, which experiences higher precipitation, richer soils, and lower evaporation rates, is a community of evergreen shrubs rather than of the drought-deciduous shrubs found farther south. Typical shrubs along the north coast are coyote brush, yerba santa *(Eriodictyon californicum),* bush monkey flower *(Mimulus aurantiacus)* with abundant poison oak *(Toxicodendron diversilobum)* and California blackberry *(Rubus ursinus)* mixed in. In sandier areas, various bush lupines (*Lupinus* spp.), intermix with coyote brush. Some ecologists have called the chaparral communities under the influence of summer fog "maritime chaparral," a fitting name for the coastal scrub communities, where fog is so prevalent.

At Palomarin, in Marin County, the Point Reyes Bird Observatory has conducted long-term studies of coastal chaparral bird communities. Among the most characteristic species of the coyote brush habitat at Palomarin are California Quail, Nuttall's White-crowned Sparrow, Wrentit, Bewick's Wren *(Thryomanes bewickii)*, and California and Spotted Towhee. In Monterey County, where the chaparral soils are sandier, the maritime shrub communities are somewhat drier and dominated by more *sclerophyllous* vegetation, especially manzanita and ceanothus, several species of which are endemic and endangered. Birds characteristic of this more arid chaparral include those found at Palomarin (and other north coast chaparral sites), but California Thrasher is a more common and Bewick's Wren a less common member of this chaparral community. Where rocky outcrops are interspersed with California sage, Rufous-crowned Sparrow *(Aimophila ruficeps)* may be present.

Farther south is the Californian coastal sage scrub community, found most abundantly on the cismontane slopes of the Transverse and Peninsular Ranges south of Santa Barbara. This is a desertlike environment, and, in fact, the ranges of several desert bird species extend coastward into the sagebrush habitat. Birds of the coastal sage scrub include species most often associated with the desert, such as Greater Roadrunner *(Geococcyx cal-*

Plate 66. Bewick's Wrens are opportunistic nesters. Here one carries an insect to its young in a discarded cardboard container.

ifornianus), Gambel's Quail, Cactus Wren *(Campylorhynchus brunneicapillus)*, and Black-throated Sparrow, as well as typical chaparral species such as California Quail, Bushtit *(Psaltriparus minimus)*, California Towhee, and Rufous-crowned Sparrow.

On south-facing slopes, almost all the shrubs are drought-deciduous, losing most of their foliage when evapotranspiration rates exceed available moisture. California sage is the emblematic shrub species here, an indicator of the coastal sage scrub community. Several other common shrubs are "sages" (*Salvia* spp.), buckwheat *(Eriogonum fasciculatum)*, coast brittlebrush *(Encelia californica)*, and lemonadeberry *(Rhus integrifolia); Opuntia* cacti (cholla and prickly pear) are common community members as well. On north-facing slopes, where temperatures are lower and moisture is more available, the coastal sage scrub community is joined by familiar broadleafed evergreen shrubs such as manzanitas, ceanothus, and toyon *(Heteromeles arbutifolia)*. Toyon bears red berries in winter and attracts frugivorous birds such as Hermit Thrush *(Catharus guttatus)* and American Robin *(Turdus migratorius)* into chaparral habitats.

Californian coastal sage scrub is a unique habitat with a high degree of endemism, but geographically limited in extent and intensely threatened by urbanization. Located on the immediate coast and surrounded by the densest human population in the state (Los Angeles, Riverside, Orange and San Diego Counties), this habitat occupies some of the most valuable real estate in the world. As a result, more than 80 percent of the original native vegetation has been destroyed.

Coastal sage scrub is also occupied by the California Gnatcatcher, a small, inconspicuous, nonmigratory songbird that has been at the center of an intense and ongoing controversy that is emblematic of the environmental challenges facing California. When the southern California building boom accelerated in the 1970s, the coastal population of gnatcatcher occupying coastal sage was considered a subspecies of the Black-tailed Gnatcatcher, a common desert scrub resident. Field work and studies of variations in plumage and vocalizations among populations provided convincing evidence that the coastal gnatcatcher was, in fact, distinct from the desert one. In 1989, California Gnatcatcher was "split" from Black-tailed, and recognized as a full species by the American Ornithologists' Union. With this elevated status, the bird was listed as "threatened" under the Endangered Species Act, and conservationists used this legal status to strengthen their opposition to developments that threatened the gnatcatcher's habitat, with some success at least initially. As the science of taxonomy was enhanced by mitochondrial DNA analysis, these newer methods were used to elucidate gnatcatcher relationships and to determine if sage scrub gnatcatchers comprised an "evolutionarily significant unit," an important distinction in terms of both conservation biology and legal protection under the Endangered Species Act. The results of these analyses showed that although California Gnatcatcher is indeed distinct from Black-tailed, the sage scrub population is not distinct from the more southerly population in Baja California, where it is common, and therefore the coastal southern California population is not an "evolutionarily significant unit," that is, it is not genetically isolated. Although loss of habitat has direct negative impacts on the species, it does not diminish the genetic diversity of the population and does not, therefore, pose an imminent threat to its continued existence . . . or so the argument has been framed.

The gnatcatcher was afforded great scrutiny because it was the "flagship" species on which efforts to conserve the habitat hinged. About 100 species of plants and animals are endemic to the coastal sage scrub habitat; however, none has been as closely studied as the gnatcatcher, and all stand to suffer as the habitat is diminished. In recent years, the concept of single-species protection, which because of the language of the Endangered Species Act has driven conservation efforts, has been expanded to encompass ecosystem or community protection. Although the law has not caught up to this perspective, regulatory agencies such as the California Coastal Commission are starting to incorporate "natural community conservation" into planning efforts. There may still be hope for the species of the coastal sage scrub.

Hard chaparral tends to be farther inland and higher upslope and to occur on steeper, south-facing slopes with thinner soils than the soft shrublands of the coastal plain. It is most abundant in southern California, but hard chaparral occurs also along the coast northward even to Mendocino County where arid conditions permit. The indicator plant of hard chaparral is chamise *(Adenostoma fasciculatum),* and its physical properties typify chaparral adaptations. It has very small, leathery leaves, inconspicuous white flowers, and an extensive root system. It is also a fire-adapted shrub that sprouts readily following wildfires either by root or burl sprouting, or by seed. Other plants of the hard chaparral include several species of *Ceanothus*, also known as California lilac; this attractive shrub tends to respond with immediate vigor after a fire. Scrub oak provides some vertical structure above the otherwise low-growing shrub layer.

Several birds—Wrentit, Bushtit, Anna's Hummingbird, Western Scrub-Jay, California and Spotted Towhees, and California Quail—may occur equally in soft and hard chaparral. Others, such as Mountain Quail *(Oreortyx pictus)*, Black-chinned Sparrow *(Spizella atrogularis)*, and Common Poorwill *(Phalaenoptilus nuttallii)*, the latter particularly toward the northern part of its range, have an affinity for drier chaparral.

Riparian Woodland

Riparian forests provide cover and refuge for a host of species, but, like other wetlands, they are also among the richest habitats for birds in and of themselves. The availability of fresh water, the rampant growth of riparian vegetation, and proximity to other habitat types provide an abundance of insects, seeds, and herbaceous forage. Deciduous riparian corridors occupy the lowlands of the Coast Ranges, usually as a narrow ribbon of habitat wending toward the coast along the valley floor or canyon bottom. Where it meets the shore or empties into an estuary, the forest

---

In early October, 1995, a widespread fire at Point Reyes National Seashore burned 12,000 acres of conifer forest and coastal scrub from the ridge down to the coastal strand. We walked into the burn a day or so afterward and stood on a barren hogback overlooking blackened hillsides; all that remained of the habitat were the skeletal branches of manzanita and coyote brush *(Baccharis pilularis)*, leafless and twisted, naked trunks of bishop pines *(Pinus muricata)*. The only greenery in the landscape was a narrow ribbon of alder and willow that followed the canyon bottom, scorched only along its outer edge. We could hear bird chatter—flickers, kinglets, and jays—coming from the riparian corridor. Walking closer, we noticed flocks of Wrentits, foraging through the foliage. Wrentits are birds of chaparral and scrub, usually seen singly or in pairs; here were flocks, taking refuge in the only habitat spared the inferno.

---

usually spreads out into a broad thicket of willow (*Salix* spp.) and alder (*Alnus* spp.). Fremont and black cottonwoods *(Populus fremontii* and *P. balsamifera)* and western sycamore *(Plantanus racemosa)* are also important riparian trees. Like other lowland habitats, coastal riparian vegetation has diminished in extent, largely owing to water diversion, grazing impacts, and clearing. Riparian-dependent species such as Willow Flycatcher *(Empidonax traillii)*, Yellow-breasted Chat *(Icteria virens)*, and Yellow

Plate 67. The Orange-crowned Warbler locates its well-hidden, cup-shaped nest on the ground. It seldom displays its colorful orange crown.

Warbler, which depend on extensive thickets or multilayered gallery forest, have diminished accordingly. Willow Flycatcher and Yellow Warbler have suffered from the added problem of cowbird *brood parasitism*. Still, large tracts occur in enough locations to provide key habitat for many of California's birds, particularly insectivorous breeders and migrants. In spring and fall, the willow thickets along the coastal corridor function as important refueling stations for migrant passerines. Some thickets in the southern part of the state provide limited overwintering habitat for species such as Wilson's Warbler. Among the most productive coastal riparian habitats are, from south to north, Tijuana River Valley, Atascadero Creek and Maria Ygnacia Creek mouths, Big Sur River and Carmel River mouths, and Smith River bottoms. However, numerous willow patches scattered along the coast provide important refueling stations for migrants leapfrogging up and down the coast each spring and fall.

In early spring, the arrival and nesting cycle of the Orange-crowned Warbler *(Vermivora celata)*, announced by sweet trills in the warm morning air, coincides with the outleafing of willow thickets. Riparian habitats in southern California host breeding Black-chinned and Costa's Hummingbirds *(Archilochus alexandri* and *Calypte costae)*, which are absent from the north coast.

Coastal Conifer and Mixed Evergreen Forest

On the outer Coast Ranges, from nearly sea level upward to 2,000 feet, coast redwood forests occur from Del Norte south to Santa Cruz County. In Monterey County, an outlier population reaches up to 3,000 feet on the north slope of the Santa Lucia Mountains. Redwoods often occur in relatively homogeneous groves where summer fog is reliable. Just inland, and often upslope to 4,500 feet, Douglas-fir becomes dominant. Adjacent to and intermixed with redwoods and fir, mixed evergreen forest predominates. On steeper slopes, broadleafed trees such as tanoak *(Lithocarpus densiflora)*, madrone *(Arbutus menziesii)*, chinquapin *(Chrysolepis chrysophylla)*, and canyon live oak *(Quercus chrysolepus)* occur in mixed stands with Douglas-fir.

The shady, moist redwood forest interior does not host an abundance of birds, and some species usually common in conifer forests, such as Pileated and Hairy Woodpeckers *(Dryocopus pileatus* and *Picoides villosus)*, actually avoid the deep forest, preferring the broken margins with some openness. Winter Wren, Brown Creeper *(Certhia americana)*, Steller's Jay, and Varied Thrush *(Ixoreus naevius)* are denizens of the shadier forest, however, and Chestnut-backed Chickadee forages in the higher branches, often in mixed flocks with Golden-crowned Kinglet *(Regulus satrapa)*, and joined in winter and spring migration by Townsend's Warbler *(Dendroica townsendi)*.

Closed-cone pine *(Pinus muricata* and *P. radiata)* forests—fairly open for conifer forests, often with dense understories of broadleafed shrubs and scrub oak or tanoak—are scattered intermittently along the coastal slope at fairly low elevations from Humboldt southward and attract birds, such as Northern Saw-whet Owl *(Aegolius acadicus)*, Hairy Woodpecker, Steller's Jay, Pygmy Nuthatch *(Sitta pygmaea)*, and Pacific-slope Flycatcher, which are often associated with mountain woodlands. Closed-coned forests are fire-adapted and tend to be associated with chaparral and coastal scrub, so birds characteristic of those habitats are also present: Bewick's Wren, Hutton's Vireo, Wrentit, and Bushtit are common breeders and year-round residents. Northern Flicker *(Colaptes auratus)*, Red-breasted Nuthatch *(Sitta canadensis)*, Yellow-rumped Warbler, and Fox Sparrow join the community in winter. It should be noted that these species (ex-

Plate 68. Acorn Woodpeckers are the most gregarious of woodpeckers, living in close-knit family groups. The black band between the white and red on the top of the head identifies this bird as a female.

Plate 69. An Acorn Woodpecker granary can contain many thousands of acorns—enough to sustain a family group through the winter.

cept for Fox Sparrow) are locally present year-round, but numbers swell greatly in winter.

Farther inland, the inner North Coast Ranges are primarily oak woodlands (see below). On north-facing slopes and higher elevations, however, mixed evergreen forest occurs as well. The drier mixed forest of the inner Coast Ranges hosts breeding bird species that occur on the outer coast, but less commonly. These dry-forest birds include Western Screech-Owl *(Megascops kennicottii)*, Ash-throated Flycatcher *(Myiarchus cinerascens)*, Cassin's Vireo *(Vireo cassinii)*, Black-throated Gray Warbler *(Dendroica nigrescens)*, Lazuli Bunting *(Passerina amoena)*, Lesser Goldfinch *(Carduelis psaltria)*, and Lawrence's Goldfinch *(C. lawrencei)*, among others.

Oak Woodlands

For the most part, oaks avoid the immediate coast, where wind and salt spray pose environmental challenges. Oaks tend to grow slightly inland, occupying the Mediterranean climatic zone, where nearly all the precipitation comes during winter and early spring; summer and fall are quite dry, even droughty. Of the 18 species of oak in California, half are "shrub oaks," common members of almost every chaparral community. Of the tree oaks, coast live oak *(Quercus agrifolia)* and valley oak *(Q. lobata)* are probably the two most characteristic of the Coast Ranges. Coast live oak, as the name implies, has an affinity for the coast zone and thrives in the milder climate—warmer winters and cooler summers—of the outer Coast Ranges. Valley oak is found inland from the fog zone, where the winter is cool and wet and the summer hot and dry. This inhabitant of the inner Coast Ranges prefers bottomlands below 2,000 feet elevation and tends to form open foothill woodlands and savannahlike forests. Both species are members of riparian communities in their respective regions.

Over 100 species of birds are associated with oak woodlands, either as breeders or erstwhile migrants. Western Scrub-Jays, in particular, are planters of acorns, and so play an active role in the dispersal and distribution of oaks. During fall, oak fruiting season, jays are busy caching acorns, usually in the ground. By one estimate, a single Scrub-Jay may stash as many as 5,000 acorns a season, but retrieves only about 30 percent. Indeed, the Miwok name for Western Scrub-Jay is "the one who plants oaks." Other *corvids* may perform this service as well—on the edge of the interior Coast Ranges, the Yellow-billed Magpie *(Pica nuttalli)*, and in the moister forests of the north coast, the Steller's Jay—yet none is as effective as the Western Scrub-Jay.

Many birds rely on oak woodlands at some time of their life cycle; however, there are several species for whom the relationship is obligatory. Acorn Woodpecker is among these oak-dependent species, rarely wandering far from a forest or savannah inhabited by oaks. Like jays, Acorn Woodpeckers harvest acorns prodigiously, but rather than burying them in the ground, the woodpeckers store their bounty in trees, "granaries," which they supply and maintain communally. The granaries, then, provide a food source for the woodpecker colony in winter and spring, long after the trees have finished fruiting. Usually, each acorn is placed in a small hole excavated for the purpose. A variety of trees are selected for granaries, but rarely do the birds use living oaks that have thin bark and hard wood. They prefer the softer, dead tissue found in thick-barked trees, or dead snags; they also use telephone poles and fence posts. In older granaries, the entire tree or post may be riddled with storage holes, but because the woodpeckers avoid living tissue, the borings tend not to damage living trees. By avoiding the moist, living tree tissue (cambium), the acorns dry more readily and are not subject to mold or rot. The colony tends their cache, rotating acorns to aid drying and moving them to smaller holes as the drying acorns shrink. Through their attentive behavior, Acorn Woodpeckers are both true stewards and true parasites of the oaks.

This stewardship of the harvest has caused Acorn Woodpeckers to evolve a complex, communal social system ("cooperative breeding"), which has been thoroughly studied by Walter Koenig and his students at the University of California's Hastings Preserve in the Carmel Valley Coast Ranges. The woodpecker colony is an extended family (kinship) group "that consists of a breeding core of four or more related males, usually brothers or a father and his sons, who share up to three related females, usually sisters or a mother and her daughters" (Koenig 1990). Older offspring may participate in the group as "helpers," attending to the granaries and assisting in raising younger chicks.

Other common birds closely allied with (though not limited to) coastal oak woodlands include Western Screech-Owl, Mourning Dove *(Zenaida macroura)*, Nuttall's Woodpecker *(Picoides nuttallii),* Anna's Hummingbird, Tree Swallow *(Tachycineta bicolor)* and Violet-green Swallow *(T. thalassina),* Oak Titmouse *(Baeolophus inornatus),* White-breasted Nuthatch *(Sitta carolinensis),* Hutton's Vireo, and Chipping Sparrow *(Spizella passerina).* When oaks are closely associated with riparian habitat, Red-shouldered Hawk *(Buteo lineatus),* House Wren, and Orange-crowned Warbler are usually present. Wood Ducks *(Aix sponsa),* too, forage on the ground beneath oaks in close proximity to streams or ponds, and the introduced Wild Turkey *(Meleagris gallopavo)* is an oak woodland visitor. Turkeys, given their abundance and voracity, are in fact detrimental to the recruitment and regeneration of oaks, eating a vast number of acorns that might otherwise survive germination.

Grasslands

Coastal grasslands have probably increased in extent since European settlement of California, replacing coastal prairie along the immediate coast, and oak woodlands toward the interior of the Coast Ranges. Marine terraces, mesas, rolling hills, and wide valleys once vegetated with moist perennial bunchgrasses and scattered vernal pools have largely been converted to annual grasslands. Grazing pressure has changed from the seasonal passage of migrating herds of antelope and elk to the year-round intensity of sedentary cattle. More recently, a conversion from annual grassland to residential and commercial development has further altered the coastal lowlands. The combined effects of ranching and urbanization have created a drier substrate and, from a bird's perspective, a less heterogeneous environment. The abundance and diversity of grassland birds has declined accordingly. Species that were adapted to the variety of microhabitats offered by those original coastal prairie-grasslands include Short-eared Owl, Burrowing Owl, Loggerhead Shrike, Western Bluebird, Horned Lark *(Eremophila alpestris),* Savannah Sparrow, Grasshopper Sparrow *(Ammodramus savannarum),* and Western Meadowlark. Of these, perhaps the Western Meadowlark and Savannah Sparrow have fared the best, but the others have each suffered declines concomitant with habitat loss.

Short-eared and Burrowing Owls are largely extirpated from the Coast Ranges, but a few still winter where appropriate habitat is available. Some grasslands still support viable populations of grassland birds. Along the north coast, just north of Crescent City, Del Norte County still has some functional coastal grasslands that support a small population of Vesper Sparrow *(Pooecetes gramineus)*. Along the central coast, where the coastal zone is protected from dense development, especially in Sonoma and Marin Counties, Savannah Sparrow and Grasshopper Sparrow are still relatively common in grasslands and prairie that are not too heavily grazed or that remain uncultivated. This circumstance is enhanced by the large amount of land protected by state parks, national parks and seashores, and the U.S. military.

Coastal Montane Communities

The majority of habitats within the coastal zone occur between sea level and about 4,000 feet in elevation. Above that, chaparral or mixed evergreen forests may give way to yellow pine *(P. ponderosa* and *P. jeffreyi)*, white and red fir *(Abies concolor* and *A. magnifica)*, or lodgepole pine *(Pinus contorta* subsp. *murrayana)* associations, supporting plants and animals very similar to those found in the montane communities of the Sierras and Cascades, but otherwise rare in the Coast Ranges. The spotty coastward distribution of Flammulated Owl *(Otus flammeolus)*, White-headed Woodpecker *(Picoides albolarvatus)*, Mountain Chickadee, Dusky Flycatcher *(Empidonax oberholseri)*, and Cassin's Finch *(Carpodacus cassinii)*, among others, shadows the distribution of these high-elevation islands of habitat where they occur within the coastal influence. Representative montane islands within the Coast Ranges are fairly common and clustered in the Klamath province and the inner Coast Ranges south to Tehama County, but become scarcer farther south. Some striking examples of montane habitat among the Coast Ranges include the Klamath Mountains, the Santa Lucia Mountains of Monterey, San Benito Mountain in the interior Coast Ranges, Big Pine Mountain in Santa Barbara County, and, of course, the San Bernardino and San Jacinto Mountains.

In the South Coast Ranges, conifer and mixed evergreen forests occur at higher elevations, or at lower elevations in drier associations intermixed with oak woodland. This transition zone (between oak woodland and boreal forest) begins on the cismontane slopes of the San Bernardino Mountains, for example, above 6,500 feet and extends upward to 9,000 feet, supporting open continuous forest of mostly yellow pines with sugar pine *(Pinus lambertiana),* white fir *(Abies concolor),* and black oak *(Quercus kelloggii)* commonly interspersed. Manzanita is a common understory shrub. Typical birds found in mountain conifers of the South Coast Ranges are similar to those of the Sierra Nevada and the montane reaches of the Santa Lucia Mountains in Monterey County: Mountain Quail, Rufous Hummingbird (*Selasphorus rufus,* a summer migrant), Hairy Woodpecker, Western Wood-Pewee *(Contopus sordidulus),* Dusky Flycatcher, Mountain Chickadee, Pygmy Nuthatch, Yellow-rumped Warbler, and Dark-eyed Junco *(Junco hyemalis).* Purple Martin *(Progne subis)* formerly belonged to this community, but has largely been extirpated from the southern California mountains. White-headed Woodpecker, although common in the Sierra and southern mountains, is absent from the Santa Lucias.

In the North Coast Ranges, at higher elevations, Douglas-fir forest blends with yellow pine (3,000 to 6,000 feet), white and red fir (above 6,000 feet), and lodgepole pine (above 8,000 feet). These forests are restricted mostly to the Klamath province. The bird community differs little from that found in the same habitats in the Sierra Nevada and Cascade Range, and so is treated in "Birds of Mountains and Foothills."

Some of the coastal peaks that reach above 4,000 feet in elevation have limited montane habitat, not enough area to "hold" mountain species, but enough to attract them during migratory periods. These vagabonds are pioneers, wandering around looking for appropriate alternatives to the usual haunts of their species. Such adventurism may come at a high cost to the individual, but as environmental conditions change and new opportunities arise, these pioneers may become "founders" of a future range extension. These mountain vagrants make for exciting possibilities when birding the upper reaches of San Francisco Bay Area peaks such as Mount Tamalpais (2,571 feet), Mount Diablo (3,849 feet), or Mount Hamilton (4,212 feet). Each has attracted Cassin's Finch, Mountain Bluebird *(Sialia currucoides),* Townsend's Solitaire *(Myadestes townsendi),* and even an occasional Northern Goshawk *(Accipiter gentilis).*

Montane islands are "migrant" traps that attract transient birds; because of often difficult access, these places are not frequently birded and would probably hold surprises for the curious naturalist.

Other Vagrant Traps

Migrating birds tend to follow distinct topographical features, especially when the celestial signposts, such as the sun and stars, are obscured by inclement weather or a thick blanket of fog. The coastline is perhaps the most prominent topographical feature, and some species habitually follow it southward en route to their wintering grounds, or northward en route to breeding grounds. Also, birds migrating or wandering westward confront the daunting Pacific Ocean when they arrive at the coast, and so are often hesitant to continue on a westward tack. For migrating birds, there is a distinct "coastal effect," with many species moving through patches of hospitable habitat during their migratory journey. Islands of vegetation—cypress trees, willow thickets, clusters of flowering vines and shrubs, or riparian forests—along the immediate coast can be hot spots for migrant birds during spring and fall. Usually the birds seen in these migrant traps are the common western migrants—Ruby-crowned Kinglets *(Regulus calendula)* and Golden-crowned Kinglets, Yellow-rumped Warblers, Hermit and Swainson's Thrushes, goldfinches, Black-headed Grosbeaks *(Pheucticus melanocephalus)*, and the "crowned" sparrows. Occasionally, an unusual bird is found, an off-course species that is rare or extremely rare in California or along the coast. The gold ring of the rare birder, however, is a "vagrant wave," a day on which all the variables—overcast skies, little wind, and timing converge, and the willows are "dripping" with out-of-range birds refueling on their quixotic journey to parts unknown.

"Rare birding" has become a favorite pursuit of many west coast birders, and the migrant traps along the outer coast, places such as the Carmel River mouth, the outer Point Reyes cypresses, the Smith River bottoms, and the Atascadero Willows, are scoured almost daily by birders, especially from mid-August into October, for vagrant warblers, orioles, or sparrows.

The subject of vagrancy is too large for this book, but the following quote on the Farallon Islands, perhaps the premier vagrant trap in North America, may give the reader a feeling for the nature of the phenomenon and the excitement it inspires in an intrepid birder: "Perhaps no other location on this continent could boast of species with such widely divergent geographical origins as Connecticut Warbler, Golden-cheeked Warbler, and Red-throated Pipit" (DeSante and Ainley 1980).

BIRDS OF THE CENTRAL VALLEY AND DELTA

In the Central Valley of California there are only two seasons—spring and summer. The spring begins with the first rainstorm, which usually falls in November. In a few months the wonderful flowery vegetation is in full bloom and by the end of May it is dead and dry and crisp, as if every plant had been roasted in an oven.

JOHN MUIR

Like the Central and South Coast Ranges, the Central Valley has a Mediterranean climate, with most of the precipitation falling in winter, from November through March. Indeed, some ecological treatments (e.g., Manolis 2003) place the Central Valley in the "California bioregion" with the South Coast Ranges because of the similarity of their seasons. The amount of precipitation varies tremendously on an annual scale (between 30 and 200 percent of average) and also on a decadal scale. As on the coast, precipitation varies widely from north to south with more than 35 inches a year at Redding and only six inches a year at

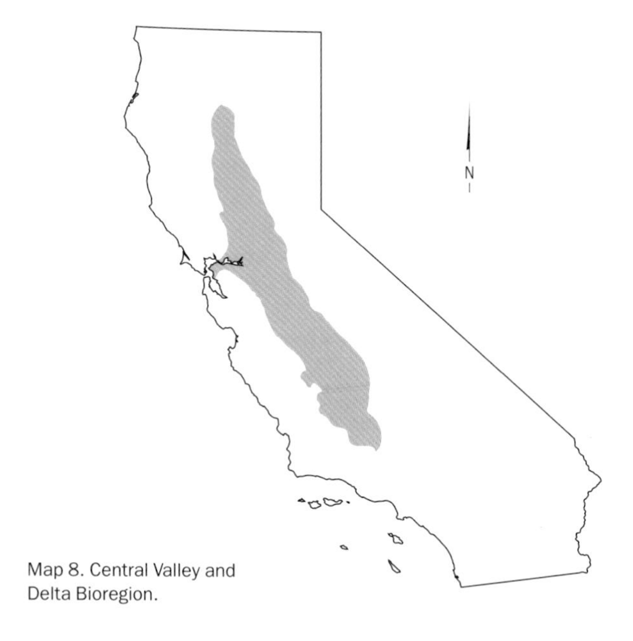

Map 8. Central Valley and Delta Bioregion.

Bakersfield. This wide variation in precipitation determines the amount of open water available to waterbirds, of course, and influences the vigor and growth of the vegetation, thereby affecting landbirds.

Topography and Ecological Context

California's Great Central Valley is a vast (60,000 square mile) basin that accounts for about one-third of California's land area. Bound by the Klamath Mountain ranges to the north, the Sierra Nevada and Cascade Range to the east, the Tehachapi Mountains to the south, and the Coast Ranges to the west. In this chapter, the Central Valley includes the adjacent slopes up to about the 1,000-foot contour, mostly dry grassy hillsides and oak woodland savannahs.

The Central Valley once supported a rich diversity of grassland-prairie, freshwater marsh, riparian woodlands, and valley oak savannah, but since the late nineteenth century, it has been converted into one of the most productive agricultural regions of the world. If Muir's hyperbole quoted above is true, the seasons are the only thing about the Central Valley that has not changed in the ensuing century and a half.

Actually, the Great Central Valley encompasses two valleys: the smaller, wetter Sacramento Valley to the north and the larger, drier San Joaquin Valley to the south. Each valley contains basins: large, shallow sinks that collect precipitation and runoff and communicate with one another hydrologically only during the wettest season. The San Joaquin contains two basins: the Tulare Lake Basin, formerly an extensive marshland, and the San Joaquin Basin. In the upper Sacramento Valley, the Redding Basin is perched in the upper reaches of the Sacramento River, separated from the valley lowland by the Red Bluff bench, and has upland (oak–pine woodland) characteristics. Across the Sacramento Valley floor lie a series of hydrologically discrete features—the Colusa, Butte, American, Sutter, and Yolo Basins.

Each valley is drained by a major river. The Sacramento River runs 250 miles from Red Bluff southward into San Francisco Bay, capturing the large tributaries that drain off the Sierras (the Feather and American Rivers), and the much smaller tributaries draining off the Cascades and the North Coast Ranges. The San Joaquin River runs 267 miles northward into San Francisco Bay, capturing the Yosemite drainages of the Tuolumne, Merced, and Stanislaus Rivers and, farther north, the Mokelumne, Cosumnes, and Calaveras Rivers. The lowlands at the confluence of the Sacramento and San Joaquin, where they merge and flow together into San Francisco Bay, are generally referred to as "the Delta." Both rivers are freshwater courses until they reach the Delta, where saltwater from the Pacific Ocean intermixes with freshwater runoff, creating estuarine conditions that vary widely with season, rainfall, water diversions, tidal range, and local topography. The Delta is a complex system of meanders, channels, islands, and levees, that has been highly altered to accommodate human endeavors since settlement of California by Europeans in the 1800s. Under natural conditions, the lower portions of the rivers flooded regularly in winter and spring, creating a network of distributary channels and a complex succession of riparian forests, tule islands, and fresh or brackish marshes. With conversion to agricultural and residential uses, the construction of levees for flood control and water budgeting has reduced some of the natural dynamism of the downstream system. Upstream, as humans moved onto the valley floor, the extensive riparian forests that occupied the riverbanks and floodplains were quickly thinned and removed:

> Today, this [native] vegetation has been almost entirely lost, mostly converted to agricultural production. Less than 5% of historical wetlands, 11% of vernal pools, and about 6% of the riparian zone remain in a quilt of disconnected patches too small to sustain dependent species. (Bay Institute 1998)

Even in their diminished capacity, the wetlands of the Delta and the valley still function as a network of habitats for birds that vary with season and environmental conditions, both natural and human influenced; however, the composition of the community has changed profoundly since precolonial days, and the change continues.

The invasion of natural communities by nonnative species usually results in a reduction in complexity. As the new generalists crowd out the old specialists, a gradual unraveling process begins, referred to by ecologists as "ecological relaxation." This process may be well under way in the Central Valley, but nature is tenacious and ingenious, and birds are among the most adaptable of her creations.

Perhaps most emblematic of the changes incurred by the Central Valley is the history of the Tulare Lake Basin. This large basin was once fed by runoff from the Kings, Kaweah, Tule, and Kern Rivers draining off the southern Sierra. Where they entered the lowlands, these watercourses formed broad deltas (alluvial fans) across the valley, which were thickly vegetated, forming continuous forests from the foothills down to the marshes surrounding the lakes. Although runoff varied tremendously with rainfall and snowpack, in wet years the basin lakes covered up to 800 square miles and were fringed by a wide tule marsh (Bay Institute 1998; Williamson 1853). Apparently, in wet years or periods of high flows, waters were exchanged between the San Joaquin River and Tulare Lake by a complex network of natural channels and sloughs.

> We camped near the lake [Tulare Lake], and in the morning I was awakened by a noise like a rush of a distant railroad train. I saw a large line of fluttering white in the far distance toward the lake, which represented, I found, an immense body of wild geese, whose wings and cries, as they moved from place to place, caused this kind of roaring noise . . . clouds of geese rose on all sides of

Plate 70. A mixed flock of Snow and Ross's Geese takes flight over Central Valley wetlands.

Plate 71. Greater White-fronted Geese rest and feed on the extensive agricultural lands in the Central Valley. Rice farmers here have been discouraged from burning their fields after harvesting, making these stubble areas attractive to foraging wildfowl.

> us, and at one point a long white line, shining like surf on a sea beach, showed us a great flock, extending more than two miles along the shore, which presently began to shiver and flare in the bright sunlight. (Nordoff 1873, in Dawson 1923)

Even with the changes that have ensued with the whole-scale conversion of the Central Valley, it remains a vital and important region for the birds of California and the North American continent. More than 60 percent of the Pacific Flyway waterfowl population winters in the Central Valley. The legions of Snow Geese *(Chen caerulescens)*, Ross's Geese *(C. rossii)*, and Greater White-fronted Geese *(Anser albifrons)* wintering in the Sacramento wildlife refuges represent a significant majority of their entire populations. Dabbling ducks—Northern Pintail *(Anas acuta)*, Northern Shoveler *(A. clypeata)*, wigeon, teal, Mallard *(A. platyrhynchos)*, and Gadwall *(A. strepera)*—gather in substantial flocks and find sustenance that bolsters their populations through the winter. Diving ducks—Bufflehead *(Bucephala albeola)*, scaup, Ruddy Ducks *(Oxyura jamaicensis)*, merganser, and Common Goldeneye *(Bucephala clangula)*—flock along overflowing rivers and flooded pastures as the larger rivers reach to-

Plate 72. Sandhill Cranes collect around a waterhole on the Carrizo Plain, where they congregate in large numbers during the winter.

Plate 73. Flocks of White-faced Ibis assemble in marshlands to feed. This species is thriving and expanding its range in California.

ward the Delta. Flocking, long-legged waders such as Sandhill Crane *(Grus canadensis)* and White-faced Ibis *(Plegadis chihi)* forage in flocks of thousands in flooded rice fields. Recent surveys (Point Reyes Bird Observatory and Partners in Flight) have shown the Central Valley also to be one of the most important regions in western North America for migrating and wintering shorebirds.

Winter surveys in the mid-1990s found 261,000 to 374,000 shorebirds in the Central Valley, qualifying it as one of the most important regions in interior North America (after Great Salt Lake) for wintering shorebirds. Plovers are well represented in lowland valley habitats in winter, particularly Black-bellied Plover *(Pluvialis squatarola)*, Snowy Plover *(Charadrius alexandrinus)*, Mountain Plover *(C. montanus)*, and Killdeer *(C. vociferus)*. Shorebirds such as Dunlin *(Calidris alpina)* and Least Sandpipers *(C. minutilla)*, which winter on the coast, apparently move inland in winter, especially during storm surges, as evidenced by November-to-January increases in the valley. Indeed, valley populations of Killdeer, Black-necked Stilt *(Himantopus mexicanus)*, Greater Yellowlegs *(Tringa melanoleuca)*, and Long-billed Dowitcher *(Limnodromus scolopaceus)* exceed those on the coast in all seasons, and San Francisco Bay is the only coastal site where shorebird numbers exceed those in the valley (Shuford et al. 1998).

From fall into spring, long-legged waders—stilts and avocets, curlews and Whimbrels *(Numenius phaeopus)*, and Greater Yel-

Plate 74. The Killdeer, in company with many other ground-nesting birds, has a strategy to lure potential enemies away from its nest by what is called "distraction display"—a convincing performance of feigned injury.

Plate 75. Black-necked Stilts nest in loose colonies, laying eggs in shallow depressions bordering brackish ponds, where they glean the surface of the water for small aquatic creatures on which they feed.

Plate 76. American Avocets choose nest sites on dry flats near water. The nest is a simple depression, lightly lined with vegetation. The thin, fragile-looking bill, used in a side-to-side sweeping motion during feeding, is, in fact, somewhat pliable.

lowlegs—rely on a variety of shallow valley wetland habitats: evaporation ponds, sewage ponds, vernal pools, managed wetlands, irrigation ditches, and flooded agricultural fields. Raptors congregate in the valley grasslands, cashing in on an abundance of rodents. Landbirds find winter refuge in the remaining riparian corridors and thickets, especially around the refuges and large tracts of open space such as the vast grasslands-wetlands complex around Los Banos (see "San Joaquin Valley," below). Huge flocks of blackbirds and cowbirds visit the agricultural fields and feedlots, and mixed sparrow flocks gather in grasslands and weed fields.

The value of the Great Central Valley as a critical bird habitat was appreciated, especially for ducks and geese, even before the network of national wildlife refuges was established in the 1940s. Its value as habitat for other waterbirds was underestimated until comprehensive surveys beginning in the 1970s (Manolis and Tangren 1975) and continuing into the 1990s (Shuford et al. 1998) served to document the valley's overall importance as waterbird habitat, resulting in the recognition of the Grasslands Ecological Area as a Ramsar site (a wetland of international importance) and as a Western Hemisphere Shorebird Reserve Network site. Active programs by the Nature Conservancy and local Audubon societies help protect and promote the valley's bird habitats, and the Central Valley Bird Club (www.cvbirds.org) has created a community of avid valley birders, who document information on distribution and abundance.

Given their grand scales and dynamic natures, each subregion of the valley—the Sacramento, the San Joaquin, and the Delta—has unique avian attributes.

Sacramento Valley

The Sacramento Valley, at a more northerly latitude and with porous, volcanic soils through much of its upper watershed, receives and retains more water than the San Joaquin. Agricultural practices have responded to this availability of water, and grain is the primary agricultural product. The Butte and Colusa Basins, in the upper Sacramento Valley between Chico and the Sutter Buttes, and Sutter Basin to the south, now form a complex of private agricultural fields and managed wetlands that supports tens of thousands of wintering waterbirds. For example, up to 6,000 Sandhill Cranes forage through the rice fields of the Butte Basin in the day and roost in the refuges at night. Snow, Ross's, and Greater White-fronted Geese, too, winter in the tens of thousands. The Tule Greater White-fronted Goose *(Anser albifrons elgasi)* breeds in the Cook Inlet of Alaska and winters almost exclusively in the Sacramento Valley. This form differs ecologically from other White-fronts and may represent a distinct species (American Ornithologists' Union 1998).

One of the premier California birding experiences is found, often quite by surprise, when you are standing on the levees at Gray Lodge, or Sacramento, or Colusa Wildlife Refuge, on a clear December evening with the sun setting red behind the Coast Ranges as flocks of geese, cranes, and dabbling ducks wing in from their feeding forays on the grain fields, flying low and steady, all converging on the safe haven the refuges provide. There comes a moment in the fading light, with the myriad flocks arriving—the sound and sheer power of all that life, that certainty of purpose converging on the wetlands—when all distractions disappear, and the attention of every sentient being is held by the gathering of these nations of birds. Even the Marsh Wrens *(Cistothorus palustris)* stop chattering. For a moment, you are enthralled, as humankind must always have been, with the plentitude of nature and the generosity of creation. You see in the refuges a remedy for all that is perverse in humanity's relationship with the natural world, a remedy to the acquisitiveness and shortsightedness our species so often exerts. These refuges are the product of people providing a sustainable system that includes the needs of every creature and honors the inherent wisdom of pure wildness.

In the center of the Sacramento Valley, the Cosumnes River Preserve protects 35,000 acres of vernal pools, grassland, marsh, and probably the most important remnant of valley oak *(Quercus lobata)* riparian habitat in the Sacramento Valley. Cosumnes provides critical habitat for riparian-dependent animals, especially California bird species of conservation concern such as Pacific-slope Flycatcher *(Empidonax difficilis)* and Willow Flycatcher *(E. traillii)*, Yellow Warbler *(Dendroica petechia)*, and Common Yellowthroat *(Geothlypis trichas)*. The Cosumnes is to the Sacramento as the Carrizo Plain is to the San Joaquin (see below), a reminder of what the landscape was like before European arrival. Most of the landbird species that occur in the Sacramento Valley floor can be found here.

The dry interior Coast Ranges border the valley's western edge, and intact habitat—a diverse mixture of oak woodland, chaparral, and grassland—still can be seen at places such as Del Puerto Canyon and Mines Road, near Livermore. Landbirds otherwise are rarely encountered in the valley—Greater Roadrunner *(Geococcyx californianus)*, Lewis's Woodpecker *(Melanerpes lewis)*, Costa's Hummingbird *(Calypte costae)*, Lawrence's Goldfinch *(Carduelis lawrencei)*--can be found with some reliability here.

Lake Solano and Putah Creek, also on the western edge of the Sacramento Valley, are relatively accessible to the Bay Area and the capital region alike, and have become favorite birding spots to find an overlap of valley and foothill birds as well as waterbirds. From Monticello Dam downstream, in an active morning, a birder can find Wood Duck *(Aix sponsa)*, Northern Pygmy-Owl *(Glaucidium gnoma)*, White-throated Swift *(Aeronautes saxatalis)*, Nuttall's Woodpecker *(Picoides nuttallii)*, Oak Titmouse *(Baeolophus inornatus)*, Phainopepla *(Phainopepla nitens)*, and most other valley riparian and chaparral species.

The low eastern foothills of the Sacramento Valley are predominantly rolling grassland and oak savannah, with stock ponds and small reservoirs scattered here and there. Spenceville Wildlife Management and Recreation Area, 20 miles east of Marysville in Yuba County, is an accessible example. The savannah supports breeding species of the oak woodland and grassland (see below), and the emergent wetland vegetation harbors wrens and rails. Black Rails *(Laterallus jamaicensis)*, formerly thought confined to tidal marshlands of San Francisco Bay, were found (surprisingly) at scattered low-elevation marshes in the 1980s. Tundra Swans *(Cygnus columbianus)* winter reliably nearby in the east valley irrigated rice fields known as "District 10," about 10 miles northeast of Marysville.

San Joaquin Valley

Historically, the Tulare Basin supported the largest interior wetland habitat in the western United States, a vast (800-square-mile) network of open water, fringing marsh, sloughs, and riparian forest. Before the tributary streams and rivers were dammed, the basin must have received spring and summer flows from the snowmelt of the southern Sierra Nevada, creating a large body of standing water. By late summer and early fall, the water level may have dropped, providing extensive shallow-water habitat and exposing an expansive and, probably, a biologically productive shoreline. We can imagine that the habitat was ripe for visitation by the migrant flocks of shorebirds and waterfowl arriving in the valley from the north. Although the avifauna were not well documented before the destruction of the shallow-water and marshy habitats, anecdotal accounts indicate that Tulare Lake and environs did support an impressive concentration of water and marsh birds. The marsh birds and many of the waterbirds disappeared with the lake, but some tenacious species have held on. Shorebird biologists have speculated that "Because of the vast extent of former habitat . . . [snowy] plover numbers in the Tulare Basin may have rivaled or exceeded those in any other region in the interior of California." Based on the accounts we have from the early naturalists, the same might be said for other interior nesting shorebirds, particularly American Avocet *(Recurvirostra americana)* and Black-necked Stilt.

Tule Lake provided viable nesting habitat for waterbirds into the 1930s and 1940s, when egg collectors found hundreds of eggs of Eared Grebes *(Podiceps nigricollis)* and American White Pelicans *(Pelecanus erythrorhynchos)*. During the development of the valley for commercial agriculture, groundwater has been pumped to the surface and distributed into shallow evaporation and retention ponds, mimicking, in a manner, historic conditions. Bird surveys of the entire Central Valley conducted in the early 1990s found that shorebird numbers were generally higher in the San Joaquin than in the Sacramento Valley and Delta; within the San Joaquin, numbers were highest in the Tulare Basin in August, during the peak period of shorebird migration statewide. That same study (Shuford et al. 1998) estimated about 15,000 Wilson's Phalaropes *(Phalaropus tricolor)* in the Tulare Basin in late July; the basin also supports large concentrations of American Avocets, Red-necked Phalaropes *(Phalaropus lobatus)*, small sandpipers (Western *[Calidris mauri]* and Least), and Black-necked Stilts in August. Concentrations of shorebirds (except phalaropes, which migrate farther southward) remain high through winter at various San Joaquin wetland sites with shifts among evaporation ponds, managed wetlands, and agricultural fields within a given season as water levels vary. Of all wintering shorebird species, the Black-necked Stilt has the strongest preference for the San Joaquin Valley.

The riparian habitat, discussed below, stretched across the valley before dams, diversion, and distribution of water so altered the natural hydrology of the lowlands. Still, some of the richness is preserved, especially along the lower 14-mile stretch of the South Fork of the Kern River, a cottonwood–willow forest that has been designated as a "Globally Important Bird Area" and a National Natural Landmark. Indeed, the Kern River watershed, encompassing an array of southern Sierra and valley habitats, has a bird list of 333 species, (including many rarities), an estimated two-thirds of which nest locally, a remarkable total for an inland site. Because the Kern River Preserve represents both valley and mountain habitats, this biologically diverse gemstone in discussed more fully in "Birds of Mountains and Foothills."

In the northern San Joaquin, the Grassland Ecological Area (including San Luis and Merced National Wildlife Refuges and Los Banos and Volta State Wildlife Areas, and several private duck clubs), encompassing 160,000 acres, comprises one-third of the wetlands remaining in the entire Central Valley. Wintering waterfowl are abundant, especially the dabbling ducks—Northern Pintail, Green-winged Teal *(Anas crecca),* Gadwall, Mallard, and Northern Shoveler. Geese are present, though not in the numbers found to the north in the Sacramento refuges. This area attracts thousands of wintering Sandhill Cranes and White-faced Ibis, with recent surveys exceeding 10,000 individuals of each species. The San Joaquin Basin has become a winter stronghold for both species. Boardwalks through the wildlife refuges here provide opportunities to get up close and personal with otherwise difficult-to-see marsh birds—bitterns, rails and moorhens, wrens, and sparrows. The grasslands are also a population center for the nearly endemic Tricolored Blackbird *(Agelaius tricolor),* a California "Bird Species of Special Concern" because of its declining population. Although Tricoloreds are "itinerant breeders," moving among areas to nest sequentially within a single season, the grasslands regularly support a colony of 5,000 to 10,000 breeding birds and even larger wintering flocks. Mendota Wildlife Area, in Fresno County, supports even larger concentrations of Tricoloreds.

> The colonial breeding system of the tricolored blackbird probably evolved in the Central Valley where the locations of surface waters and rich sources of insect food were ephemeral and varied annually (Orians 1961). Before its rivers were dammed and channelized, the Central Valley flooded in many years, forming a vast mosaic of seasonal wetlands, freshwater marshes, alkali flats, native grasslands, riparian forests, and oak savannahs. Virtually all these habitats once supported nesting and foraging tricolored blackbirds. (California Department of Fish and Game and Point Reyes Bird Observatory 2001)

In Stanislaus County, on the eastern side of the valley and along the San Joaquin River, the outlying country around the town of Modesto is a modest, though productive, riparian corridor with sewage ponds (596 acres) that attracts open-water species. During fall, winter, and early spring, gulls, plovers, and other shorebirds occur, along with the expected landbirds. Pacific Golden-plovers *(Pluvialis fulva)* sometimes winter here, one of the few spots in North America where they do.

Kings County, in the southwestern corner of the San Joaquin Valley, is largely agricultural valley floor with some low inner Coast Ranges slopes along the western edge. This region exemplifies the conversion of valley habitat to human uses (symbolized by vast blackbird flocks attending barnyards), and also the tenacity of birds seeking sustenance in those habitats remaining. Corcoran Reservoir and variably flooded fields attract waterbirds in the lowlands, and the western hills support breeding populations of Grasshopper Sparrow *(Ammodramus savannarum)*, Vesper Sparrow *(Pooecetes gramineus)*, Sage Sparrow *(Amphispiza belli)*, and Rufous-crowned Sparrow *(Aimophila ruficeps)*.

In the southeastern corner of the San Joaquin, three old lake beds in the Tulare Basin—Tulare, Buena Vista, and Kern—have been taken over by corporate agricultural land-use practices; viable bird habitat persists, mostly in spite of rather than because of people's efforts. Water contaminants associated with these agricultural practices, especially high concentrations of selenium, have insinuated themselves into the tissues of locally nesting shorebirds such as Black-necked Stilts (and probably others), impairing breeding success. The high concentrations of shorebirds using the shallow-water banks and flooded rice fields, especially in late summer—avocets and stilts, Red-necked Phalarope, Least and Western Sandpipers—must be picking up these alien constituents in their prey and carrying them to their breeding grounds, whether in California or the high Arctic. As John Muir said: "When we try to pick out anything by itself, we find it hitched to everything else in the Universe" (1911).

The Carrizo Plain, 45 miles long and 10 miles wide, is a bioregion unto itself, with affinities to the Coast Ranges, the desert, and the San Joaquin Valley. Because of its proximity (on the western edge) and its expanses of native grasslands, the plain is included here as part of the San Joaquin Valley. Indeed, habitat within the Carrizo Plain may more closely resemble the pristine San Joaquin than any area within the valley itself, so altered has the landscape become. Carrizo Plain National Monument (designated by President Clinton on January 17, 2001) protects 250,000 acres and includes both the Carrizo and Elkhorn Plains and Soda Lake, an alkali basin. When winter rainfall is sufficient, Soda Lake provides shallow-water habitat for thousands of waterbirds, including grand flocks of Sandhill Cranes, which dependably winter in the basin.

Carrizo's affinity with the California deserts is also expressed in its birdlife. LeConte's Thrasher *(Toxostoma lecontei),* Scott's Oriole *(Icterus parisorum),* Brewer's Sparrow *(Spizella breweri),* and Black-throated Sparrow *(Amphispiza bilineata)* all occur here. The Thrasher and Oriole reach the northern limit of their range at Carrizo Plain. The Cuyama Valley, just to the south, supports similar habitat characteristics and bird communities.

Delta

At the Sacramento–San Joaquin Delta, the two great rivers converge, delivering 90 percent of California's runoff into the state's largest and North America's most biologically diverse estuary, San Francisco Bay. The Delta encompasses nearly 800 square miles of lowlands, prone to flooding and highly variable conditions despite people's best efforts to constrain nature's impetuousness. All the species of waterbirds that occur in the Central Valley, and most of the landbirds, pass through the Delta at some stage of their lives. It is difficult to convey the energy that the Delta encompasses, or the changing complexity of its birdlife.

SANDHILL CRANE RESERVE

JANUARY 10, 2003. ISENBERG SANDHILL CRANE RESERVE, WOODBRIDGE ROAD, SAN JOAQUIN COUNTY.

Arrived at about 4:00 p.m. Sky 100 percent clear, about 45 degrees F, wind westerly less than 10 miles per hour. Mount Diablo incandescent, outlined in crimson by the setting sun. First sounds heard are the strident alarm calls of Killdeer, Black-necked Stilts, and Greater Yellowlegs, triggered as I set up my scope. Wigeons whistle, Mallards grunt, and shoveler, Gadwall, and Green-winged Teal forage in the water, undisturbed. A dozen Tundra Swan and a single American White Pelican roost in the far corner of the ponds, preening, settling in for the night. Flocks of hundreds of Greater White-fronted Geese are moving overhead, flying northward. A string of Snow Geese off to the west crosses in front of Diablo, also heading north; two smaller white geese, Ross's, are trailing. Hearing some pipits and Horned Larks behind me, I turn and see a Northern Harrier coursing low across the stubbled field. Then the cranes start to arrive, small flocks at first, calling that throaty, sonorous "craa-craa-craa-craaoooooo," that greeted the first people who arrived here 8,000 or maybe 12,000 years ago. Cranes land on the levee, one after the other, with a lightness uncommon even among the most graceful of birds. The light is fading, and I can no longer make out their crimson heads through my scope. I look up and see that hundreds are now arriving in flocks of two or three dozen, coming in from all directions, especially from the south. The evening resonates with their deep-throated, haunting, soundings. Soon it is dark, quiet and still, only the frogs are singing.

Winter, however, is the season of greatest exuberance, brought to life by the movement of flocks to and from fields and marshes, as they shift among the variable resources available. The fieldnote excerpt, "Sandhill Crane Reserve," may provide a glimpse into the Delta's vitality.

Unique attributes of the Delta include a roost of approximately 30 Swainson's Hawks *(Buteo swainsoni)*, which have wintered here annually since 1991 and are the only confirmed regularly wintering population in California. The Delta is also a favored wintering grounds for the Tule Greater White-fronted

Plate 77. Swainson's Hawks are birds of wide-open spaces, where they search for ground squirrels and other small mammals while soaring or by watching from a perch.

Goose and the Aleutian Canada Goose *(Branta canadensis leucopareia),* and, on the bench islands that border the sloughs, small numbers of breeding Black Rails. The breeding distribution of some valley birds, such as Blue Grosbeak *(Passerina caerulea),* reach their western limit in the Delta.

Habitats

Wetlands

Lakes and Flood Basins

Prior to alteration in the nineteenth century, natural flood-basin wetlands were the primary habitat of the lowland valley, existing as scattered but interconnected ponds, marshes, and pools over a distance of about 300 miles from near Chico southward to Visalia. Extensive tule and cattail marshes surrounded the standing water of the larger, deeper water bodies.

When considering waterbird species in the valley wetlands, we must acknowledge the changes that have occurred since the conversion of the valley to agricultural production. Tulare Lake was the largest body of fresh water in the American west. "In the broad waters of the lake [Tulare] lived mussels, clams, turtles, and a wide variety of fish. Most abundant were the birds, especially geese and ducks, which in autumn and spring filled the sky in numbers beyond counting" (Kirk 1994).

Within 50 years after the Gold Rush, the diversion of feeder streams for agricultural irrigation drained the lake of its water and wildlife. By 1882, the shoreline of Tulare Lake had receded by an average distance of "something like twelve miles" (*History of Fresno County, California* 1882). By 1922, Tulare Lake, once, at 800 square miles one of the major wetlands on the Pacific Flyway, had begun to vanish from the landscape. The lake persisted, intermittently, when flooding rejuvenated it. For example, in the wet winter of 1937, it was estimated to cover 30 square miles. Today "Tulare Lake" is "a series of evaporation ponds covering some 7000 acres . . . rich with poisonous pesticides and salts leached from the soils by irrigation" (Kirk 1994). The historic wetlands of the southern San Joaquin have been reduced by more than 95 percent of their former extent. But so rapidly was the basin drained, so absolute its metamorphosis, that only a fragmentary record of its early avifauna survives. For example, in the *Game Birds of California* (Grinnell et al. 1918), the authors observed: "In the San Joaquin Valley, cranes have been observed during the summer months, and there is a chance that they may breed there, or at least have done so."

But we do know some of the avifaunal history. American White Pelican, once a common nester, has been exiled from the Central Valley, at least as a breeder. Locations where American

Plate 78. The Marsh Wren is found among tall aquatic vegetation such as tules or cattails. The polygynous male will build several nests to attract the attention of females.

White Pelicans formerly bred include sloughs and sequestered islets of the lower Sacramento River (until 1910), Tulare Lake (until 1912, intermittently up to 1941), and Buena Vista Lake, Kern County (until 1923). Surely less-conspicuous species than the pelican fared no better with the draining of Tulare Lake. The extensive tule marshes around the lake's perimeter must have supported large numbers of American Bittern *(Botaurus lentiginosus)* and Least Bittern *(Ixobrychus exilis)*, Black-crowned Night-Heron *(Nycticorax nycticorax)*, Sora *(Porzana carolina)*, Virginia Rail *(Rallus limicola)*, Marsh Wrens, and Common Yellowthroats.

American Avocets and Black-necked Stilts, too, were abundant nesters on old lake-bed islands. Before the conversion of the tidal marshes of much of San Francisco Bay to salt ponds, where large numbers of both species now breed, Tulare and Buena Vista Lakes were important population centers for these long-legged waders. In the southern San Joaquin in the early twentieth century, stilts were considered much more abundant than avocets and, according to J. G. Tyler (quoted in Grinnell et al. 1918), outnumbered them 100 to 1, "and as avocets are decreasing the disparity is increasing." (A Tyler aside: "An exceptional condition . . . was observed at Buena Vista Lake, Kern Co., where Avocets and Stilts were nesting on common ground . . . indeed, some nests contained eggs of both species!") These similar species illustrate, in fact, the changing distribution patterns that California birds exhibit. In recent years, avocets have extended their breeding range northward to Humboldt Bay, and stilts, too, are increasingly common around San Francisco Bay. Stilts are still perhaps the most common nesting shorebird in the valley, and other species have increased in abundance, both in winter and summer. Increases in White-faced Ibis, California Gull *(Larus californicus)*, Black Tern *(Chlidonias niger)*, and Forster's Tern *(Sterna forsteri)* are also well documented.

The Tulare Basin still floods in wet years, and southeast Kings County still has the distinction of being identified as "the largest remaining wetland in the San Joaquin Valley." Birds respond to available habitat, not necessarily to its aesthetic or natural characteristics. Although the geese and pelicans once so abundant are gone, what are effectively ephemeral shallow reservoirs, when hydrated, still attract transient waterbirds (ducks, shorebirds, and gulls), breeding Caspian Terns *(Sterna caspia)* and

Forster's Terns and Snowy Plovers, and substantial wintering flocks of shorebirds.

Managed wetlands of the Central Valley—including national wildlife refuges managed by the U.S. Fish and Wildlife Service such as Los Banos, Colusa, and Sacramento; the California Department of Fish and Game's Gray Lodge Wildlife Refuge; flooded agricultural fields; and private hunting preserves—support impressive flocks of waterfowl, especially in winter. Snow Geese, Ross's Geese, and Greater White-fronted Geese *(Dendrocygna bicolor)*, Sandhill Cranes, and an abundance of dabbling ducks—Mallard, Gadwall, Northern Pintail, Green-winged Teal, and Cinnamon Teal *(Anas cyanoptera)*—congregate in refuges and adjacent grain fields. Indeed, the statement made by Grinnell et al. (1918) in reference to the Snow Goose, that "during the winter months almost the entire population concentrates in central and western California" is still true. However, one authority estimated that 60 million waterfowl once visited the Central Valley; now the estimates are at about 2.5 million birds. (Since the early twentieth century, the Klamath Basin has been developed as an area that now also supports those wintering flocks and augments the Central Valley wetlands. See "Birds of the Great Basin").

Plate 79. A Wilson's Snipe keeps vigilant watch over the wet meadow in which it nests.

Vernal Pools

Vernal pools are shallow depressions underlain by soils that drain slowly (or not at all) and therefore retain runoff and rainwater during winter into early spring. Before conversion of the landscape, vernal pools were a common characteristic of the valley, a unique wetland type that must have provided extensive habitat,

Plate 80. A male Tricolored Blackbird displays distinctive red-and-white "epaulets." This gregarious bird nests in large colonies in the lowlands and foothills.

especially for early spring migrants such as Blue-winged Teal *(Anas discors)* and Greater Yellowlegs.

Vernal pools provide critical links as birds move between and among wetland habitats at both local, regional, and continental scales. Although they originally comprised 415,000 acres in the Central Valley, only about 10 percent remain. Recent preservation and even restoration efforts, however, are increasing their extent, recreating some semblance of the pristine Central Valley environment. When flooded in late winter and early spring, these productive ponds offer optimal foraging depths for some waterbirds—dabbling ducks (teal, shoveler, wigeon), waders with relatively long legs (stilts, avocets, yellowlegs, and dowitchers)—all refueling for the imminent migration or nesting phase of their lives. Black-necked Stilts often breed in the wet prairie meadows associated with vernal pools. Adjacent moist meadows and grasslands provide habitat for Wilson's Snipe *(Gallinago delicata)*, Grasshopper Sparrow, and Western Meadowlark *(Sturnella neglecta)*. The barren ground of dry pools are used by a variety of passerines such as American Pipit *(Anthus rubescens)*, Horned Lark *(Eremophila alpestris)*, Tricolored Blackbird, and Brewer's Blackbird *(Euphagus cyanocephalus)*, and even as nesting sites by Lesser Nighthawk *(Chordeiles acutipennis)*. Some vernal pools are now protected at places such as Jepson Prairie Preserve, Solano County. At other locations, they are subject to the usual threats of development and water diversions. Some important vernal pool habitats whose fate remains in question include those at Willow Slough, Madera County, the Merced Grasslands just north of Fresno, the grasslands between the Tuolumne and Stanislaus Rivers (Cooperstown/Red Hills area) at the western edge of the Sierra foothills, and Rancho Murieta Grasslands east of the Cosumnes River Preserve near Sacramento.

The Riparian Zone

> *In their pristine condition, the streams of the Sacramento river system were flanked by forests . . . where the natural levees are widest, the riparian forests achieved their greatest width, four to five miles.*
>
> K. THOMPSON, *RIPARIAN FORESTS OF THE SACRAMENTO VALLEY*

Estimates vary on the original extent of riparian acreage in the Central Valley; however, one million acres falls within the middle range of most attempts: 637,000 acres in the Sacramento Valley; 329,000 in the San Joaquin; and 42,000 in the Delta. These extensive forests were decimated in the nineteenth century; only an estimated 10 percent remains. Not until the late twentieth century did we, as a society, begin to understand the value of the riparian zone to water quality and wildlife. Consequently, protective measures, restoration efforts, and a system of preserves have now been established in an effort to reverse, or at least halt, the destruction of this critical habitat.

The density of vegetation in the riparian zone depends on the amount of moisture available in the soil, which generally decreases with distance from the watercourse. So, closest to the river and on lower, frequently flooded banks and terraces, riparian vegetation, usually a mixture of willows and cottonwoods, forms a dense band of habitat with a well-developed canopy layer that runs parallel to the watercourse. Sand- and gravelbars within the watercourse are readily colonized by willows. Beneath the canopy, a luxuriant understory develops when grazing pressure is absent and flooding is infrequent. These riparian understory thickets—relatively moist, cool, and shaded—provide an important microclimate for nesting and migratory birds. Even species that nest and forage out in the drier, more *xeric* portions of the valley floor visit the riparian zone for water and refuge during some portion of their life cycle. The diversity and abundance of birds (and other wildlife species) is greatest, in fact, where riparian and upland habitats occur together or in close proximity to each other. In fact, access to riparian habitat is critical to most wildlife species and provides a vital link to the broad biodiversity of any region.

The undergrowth of a riparian strip closest to the water (the "mesoriparian zone"), when healthy, has a high percentage of woody vines (lianas), including blackberries (*Rubus* spp.), which account for nearly half the ground cover. Shrubby trees and bushes such as buttonbush *(Cephalanthus occidentalis),* Oregon ash *(Fraxinus latifolia),* and blue elderberry *(Sambucus mexicana)* are also common.

Breeding species of the Central Valley riparian zone have suffered significant declines with the elimination of broad swaths of vegetation, especially willow and cottonwood forests. Neotropical insectivores such as Willow Flycatcher, Warbling Vireo *(Vireo gilvus),* Yellow Warbler, and Least Bell's Vireo *(Vireo belli pusillus)* have been all but eliminated as breeding species from their former haunts. Not only has their preferred breeding substrate and foraging habitat been removed, but also in those habitat patches that remain nests are more prone to predation for lack of cover. Concurrent with the diminution of riparian vegetation, populations of potential nest predators have increased, subsidized in part by agricultural cultivation and grains available at feed lots. Corvids—scrub-jays, crows, and magpies—undoubtedly take their toll on eggs and nestlings; however, the culprit most often identified as responsible for the decimation of the smaller songbirds is the Brown-headed Cowbird *(Molothrus ater).* As early as 1944, Grinnell and Miller noted the decline: "In the last fifteen years a noticeable decline in numbers [of Least Bell's Vireo] has occurred in parts of southern California and in the Sacramento and San Joaquin Valleys, apparently coincident with increase of cowbirds which heavily parasitize this vireo."

Other Central Valley riparian or marsh-edge passerines have faired better, so far. Blue Grosbeak and Song Sparrow *(Melospiza melodia),* perhaps because of their coarser habitat requirements, larger sizes, and more aggressive natures, still occur widely, if in reduced numbers, throughout the valley.

The remnant riparian forests of the Central Valley now have

diminished capacity for many native breeding birds; however, their value for migrant songbirds has perhaps been amplified by their scarcity. In one fall migration study by Point Reyes Bird Observatory biologists, 125 landbird species were recorded in valley riparian thickets; nearly one-quarter of those species were *neotropical migrants;* the rest were fairly evenly divided between residents and winter visitors. In terms of numbers of individual birds, however, neotropical migrants outnumbered residents or winter visitors. The five most common species captured in mist nets were White-crowned Sparrow *(Zonotrichia leucophrys),* Yellow Warbler, American Goldfinch *(Carduelis tristis),* Ruby-crowned Kinglet *(Regulus calendula),* and Orange-crowned Warbler *(Vermivora celata).*

Oak Woodlands and Savannah

Forests of valley oaks, whose roots need year-round moisture, once bordered most of interior northern California rivers and wetlands, and were essentially unaltered by humans until 1849 (Thompson 1980). Originally, the outer edge of the riparian zone (the "xeroriparian") graded into valley oak woodlands with widely spaced grasslands between, creating a savannahlike landscape. Many small seasonal creeks traversed the upper floodplain, joining major watercourses in the valley bottom. This oak savannah was more extensive in the Sacramento Valley than in the San Joaquin, covering nearly 1.5 million acres.

Above the valley floor on the lower foothills, valley oak gives way to the more drought-tolerant blue oaks *(Quercus douglasii),* and savannah gives way to a forest community that includes foothill pine *(Pinus sabiniana),* buckeye *(Aesculus californica),* and other oaks (interior, coast, and canyon live oaks *[Q. wislizenii, Q. agrifolia,* and *Q. chrysolepus]*). On the slopes above the valley floor, the trees are closer together and form more of a canopy, providing more cover, more cavities, and more structural diversity for nesting birds than on the open savannah. The variety of birds (and other animals) is also higher here than in adjacent habitats—grasslands, chaparral, and conifer forest.

The mixed oak woodlands of the foothills also attract mixed flocks of migrant insectivores in fall and spring, rivaling riparian habitats for diversity and avian activity. As energetic migrant flocks of flycatchers, warblers, vireos, and kinglets pass through

Plate 81. Western Bluebirds prefer open woodland. They nest in tree cavities or in suitable nooks in buildings, and they are often attracted to nest boxes positioned on a fence running through grassland habitat.

the foliage, they excite and attract the resident titmice and bushtits, creating a flurry of activity. The concentrations of small species in these mixed flocks can be impressive, and the curious naturalist becomes enthralled finding neotropical migrants such as Black-throated Gray Warbler *(Dendroica nigrescens)*, Townsend's Warbler *(D. townsendi)*, Orange-crowned Warbler, and Nashville Warbler *(Vermivora ruficapilla)* moving through the greenery with forest birds—woodpeckers, thrushes, kinglets—that moved downslope from higher elevations in fall.

About 75 species nest in the oak woodlands of the valley and adjacent foothills. Although many nesting species move around locally among habitats, year-round residents include raptors such as Red-tailed Hawk *(Buteo jamaicensis)* and Red-shouldered Hawk *(Buteo lineatus)*, Great Horned Owl *(Bubo virginianus)*, Western Screech-Owl *(Megascops kennicottii)*, and Northern Pygmy-Owl, California Quail *(Callipepla californica)*, Mourning Dove *(Zenaida macroura)*, Yellow-billed Magpie *(Pica nuttalli)*, and Western Scrub-Jay *(Aphelocoma californica)*, Nuttall's Woodpecker and Acorn Woodpecker *(Melanerpes formicivorus)*, Northern Flicker *(Colaptes auratus)*, Anna's Hummingbird *(Calypte anna)*, Oak Titmouse, White-breasted Nuthatch *(Sitta carolinen-*

Plate 82. The Black-throated Gray Warbler is a spring and fall migrant throughout the Central Valley.

Plate 83. The White-breasted Nuthatch is the largest of the three species of North American nuthatches. Nuthatches tend to feed while descending the bark of trees, a tactic that possibly enables them to see insects, spiders, and other prey missed by woodpeckers and creepers that ascend trees to forage.

Plate 84. Small groups of Lewis's Woodpeckers sometimes choose to nest in the gaunt snags of burnt-out forests. This woodpecker lacks the characteristic undulating flight pattern of other woodpeckers.

Plate 85. Like all members of its family, the Black-chinned Hummingbird is an aerial acrobat. Black-chins are rare on the coast and in high-mountain regions but common in the interior lowlands.

sis), Bushtit *(Psaltriparus minimus),* Hutton's Vireo *(Vireo huttoni),* California Towhee *(Pipilo crissalis),* and American Goldfinch.

Wanderers such as Phainopepla and Cedar Waxwing *(Bombycilla cedrorum)* move around the valley perimeter seasonally, but may be found in winter in the oak canopy, when and where mistletoe berries are available. Though more sedentary, Western Bluebirds *(Sialia mexicana)* and Northern Mockingbirds *(Mimus polyglottos)* are also attracted to the mistletoe.

Characteristic oak woodland summer residents, species that arrive in spring and migrate southward in fall, include Ash-throated and Pacific-slope Flycatchers, Black-headed Grosbeak *(Pheucticus melanocephalus),* Orange-crowned, Black-throated Gray, and Nashville Warblers, Bullock's Oriole *(Icterus bullockii),* Tree Swallow *(Tachycineta bicolor),* and Violet-green Swallow *(T. thalassina),* among others.

Grasslands

The California grassland province of the Central Valley once occupied about 20,200 square miles (52,300 km^2) in a flat alluvial valley between the Coast Ranges to the west and the Sierra Nevada to the east. Regions of the valley north of Fresno that receive 10 inches or more of rainfall per year were dominated historically by bunchgrasses similar to those of the Great Plains. The original grasses have been mostly replaced by introduced annuals, predominantly European weeds. Perennial bunchgrasses capture and hold air moisture more tenaciously than the annuals, and, being viable all year, protect the soil from solar radiation. Annuals die out during the driest months, exposing the soil to more solar radiation. This wholesale ecological conversion has caused an overall desiccation of the soil, affecting water retention capacity and runoff rates, and, ultimately, these habitat effects extend beyond the reach of the grasslands.

Steppelike grasslands tend to hold relatively few species that fluctuate widely in abundance seasonally. California's grassland communities have affinities with those of the Great Basin, with most bird species occurring in both regions. For example, Ferruginous Hawk *(Buteo regalis),* Long-billed Curlew *(Numenius americanus),* and Mountain Plover winter in both areas; Burrowing Owl *(Athene cunicularia),* Western Meadowlark, and Horned

Lark breed in both areas. It is worth noting that Mountain Plover and Long-billed Curlew—both steppe-associated shorebirds that winter in the valley—exemplify the diversity that has evolved within the shorebird group, both morphologically and ecologically. Both species are grassland specialists; plovers favor dry ground and curlews tend to occur on moister substrate. Dramatic differences in the bills of each species have evolved to exploit their respective habitats.

Typical of open country, grassland birds are to be expected in the Central Valley. Burrowing Owl, Short-eared Owl *(Asio flammeus)*, and Grasshopper Sparrow are breeding species whose populations have been decimated by habitat conversion of the valley floor. These species still hold on at a few outlying locations such as the Merced Grasslands of eastern Merced County. Vesper Sparrow is a winter visitor, though sparingly, to dry grasslands of the San Joaquin Valley and the Carrizo Plain, preferring flat or gently sloping areas with some open ground. Any doubt about habitat preference is dispelled by this species' scientific name: *Pooecetes*, "grass-dweller," and *gramineus*, "fond of grass."

The valley grasslands, plains, and agricultural fields provide particularly good habitat for raptors. Two *buteos*, Ferruginous

Plate 86. The Long-billed Curlew has a bill adapted to probing deep into the mud of tidal flats and lagoon edges for burrowing crustaceans. It forsakes this habitat to breed in the dry upland meadows of northeastern California, where it feeds mostly on insects.

Plate 87. Two young Red-tailed Hawks, almost ready to leave their cliff-side nest, await the arrival of a parent with food.

and the Swainson's Hawks, are two valley grassland species that partition out the habitat. Ferruginous Hawk, being the largest buteo, tends to dominate other congeners (Red-tailed Hawk, Swainson's Hawk, Rough-legged Hawk *[Buteo lagopus]*), is more solitary, and takes primarily ground squirrels and rabbits. Swainson's Hawk is smaller, more gregarious (especially in migration), and takes smaller rodents as well as some birds and insects. Swainson's migrates to South America in winter, often in flocks, whereas Ferruginous stays in North America. Ferruginous arrives in the valley from its more northerly breeding grounds in fall, after the Swainson's has left for its southern wintering grounds. The only time these two species overlap is in spring, after Swainson's has returned and before Ferruginous has left. Spring is also the season of greatest rodent abundance, when the habitat is able to accommodate the appetites of two (or more) potential competitors.

Large, broad-winged raptors seen circling over the Central Valley flatlands in summer, often several individuals together, are likely Swainson's Hawks. Their flight profile is reminiscent of that of a Turkey Vulture *(Cathartes aura)* (also common), but most Swainson's are light underneath, with dark flight feathers and light wing linings, a pattern unlike that of vultures or other buteos.

Dark-phase Swainson's are rarer, and pose more of an identification problem. Swainson's have declined throughout California and the southwest over the past several decades; however, important nesting concentrations of Swainson's Hawks occur along Willow Slough, just west of Sacramento, Yolo County, and in the northern San Joaquin Valley, at the Grasslands Ecological Area, Merced County, and along the Stanislaus River near Ripon.

Another raptor, the White-tailed Kite *(Elanus leucurus)*, is a "mictrotine" obligate, meaning that it preys primarily on small rodents (especially the California Vole *[Microtus californicus]*) that eat grass herbage and seed. In years of average rainfall, especially when there is substantial precipitation in spring, grasses thrive and vole populations respond accordingly, producing a large number of young. When rodent populations boom, kites respond in kind with higher reproductive success. The relationship between rainfall and rodent populations is not always obvious, however. During drought periods, annual grasses have a long-term strategy for overcoming the lack of water. Rather than putting their energy into vegetative growth, they produce an abundance of seeds, thereby investing in future generations rather than in their own vigor. When an abundance of seed is available, *granivorous* rodents get fat, of course, and breed abundantly. The rodents, in turn, benefit the kite population. Extreme rainfall, or lack of spring rain, however, may suppress plant growth and seed production and cause a "bust" in the rodent population. Kites and other damp-grassland-loving raptors such as the Short-eared Owl have developed a strategy to deal with the uncertainty of these boom-and-bust cycles of microtine populations and the sporadic availability of their prey at varied locations. Kites are highly nomadic, wandering across the valley or into the Coast Ranges and foothills in search of rodent blooms. When they find a reliable food source, they may establish a nest site and remain there to breed. If there is an abundance of rodents, several pairs of kites, or sometimes many pairs, may establish a temporary colony and forage cooperatively through the same meadow, marsh, or fence row.

Saltbush and Alkali Sink Scrub

Several saltbush shrubs of the genus *Atriplex* are characteristic of this plant community: seep weed (*Suaeda* spp.) and iodine bush

(*Allenrolfia* spp.) are common associates. This is a desertlike environment with low annual rainfall, poorly drained soils, and sparse vegetation, usually widely spaced with open ground in between. Saltbush scrub still exists where it has escaped the plow and grader, mostly on dry hillsides in remote regions at the valley's edge, as in the southern San Joaquin in the Tulare Basin or the adjacent Kern National Wildlife Refuge. The latter, a large (10,000-acre) tract of alkali salt scrub and freshwater marsh, protects important natural valley habitat, including intact scrublands. Efforts to restore the natural characteristics of the refuge by elimination of cattle grazing have allowed species otherwise rare in the Central Valley to remain: Short-eared and Burrowing Owls, Greater Roadrunner, and Sage Sparrow. Wintering Mountain Plovers may sometimes be found in this habitat and associated dry grasslands, but that species' distribution seems to be shifting from the Central Valley to the Imperial Valley. In the Temblor Range of the western San Joaquin Valley, the alkali scrub community supports Le Conte's Thrasher, which also is associated with desertlike grassland washes in the Carrizo Plain.

BIRDS OF MOUNTAINS AND FOOTHILLS

When Theodore Roosevelt first met John Muir, it is reported that his first question was, "How can one tell the Hammond Flycatcher from the Wright Flycatcher?" It is doubtful whether any future President of the United States will have the slightest interest in the problem.

RALPH HOFFMANN, *BIRDS OF THE PACIFIC STATES*

John of the Mountain's answer to the Bull Moose is lost to history, and the Wright Flycatcher is now called the Dusky Flycatcher *(Empidonax oberholseri),* but Hoffmann's prediction was nearly prescient. The only birder among subsequent Presidents was Jimmy Carter. Field ornithologists have since refined the identification of these two nearly identical-looking *Empidonax* flycatchers, using subtle characteristics such as wing length relative to tail length, differences in call notes, and behavioral quirks of each. Given Muir's attentiveness to the subtleties of the natural world, his answer may have been similar to the observations of David Gaines (1977) in his classic *Birds of the Yosemite Sierra:* "Hammond's are wedded to the deeply shaded foliage under-

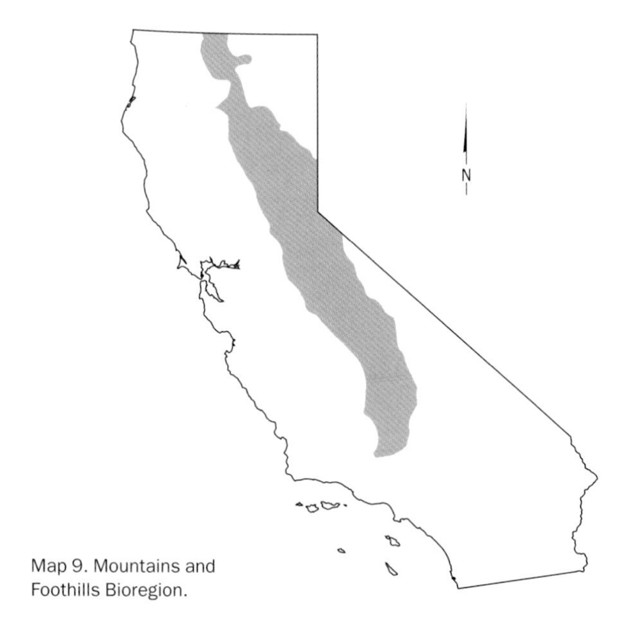

Map 9. Mountains and Foothills Bioregion.

Plate 88. Dusky Flycatchers inhabit open, high-elevation pine forest, where a pair may build a nest, often unconcealed, in the fork of a small tree. These exposed nests are often parasitized by cowbirds.

neath the crowns of lofty conifers. Duskys, on the other hand, prefer open forest edge and shrub-covered slopes where sun penetration is relatively high."

The Mountain Provinces

The mountains of California are dominated by the southern Cascade Range and the Sierra Nevada, which form the backbone of the state. The southern Cascade Range is a moderately high-elevation chain of volcanic mountains that trend southward from Oregon's Columbia Gorge to the California border, where the Klamath River valley bisects the mountains. Eighty miles farther south, the Pit River also interrupts the cordillera. From the California border south, the Cascades, blocked from the moist Pacific air by the Klamath Mountains, are covered by a relatively dry, open montane forest before merging with the granitic Sierra Nevada at the headwaters of the Feather River. Mount Shasta (14,162 feet) and Mount Lassen (10,466 feet) are two dramatic topographical features of the southern Cascades, which are otherwise modest mountains. California's Cascades, including Shasta and Lassen, are in the rain shadow of the Klamath Mountains to the west, and so are considerably drier than Oregon's Cascades. The moist Pacific air again influences the montane environment along the western slope of the Sierra, however.

The Sierra Nevada, a massive granitic, tilted-fault-block mountain range, reaches nearly 400 miles from the southern Cascades to the Transverse Ranges. It averages about 60 miles in width, covering 20,000 square miles, with most of its surface area on the western slope, which inclines gradually and carries a dozen large rivers to the Pacific. The crest is near the eastern face, a precipitous fault scarp that drops abruptly to the Great Basin below. The northern third of the Sierra rises to moderate elevations, but the southern two-thirds includes the rugged alpine peaks of the High Sierra, a region mostly above 10,000 feet in elevation extending from Sonora Pass (9,626 feet) south to Olancha Peak (12,136 feet) above the Kern River valley and including Mount Whitney (14,496 feet), the highest of 11 Sierran peaks over 14,000 feet. The western slope receives precipitation from the Pacific, most of which comes as snow at higher elevations. The eastern slope lies in the rain shadow of the crest and receives less than half as much precipitation. Deprived of the moderating influence of moist Pacific air, the eastern slope is much colder in winter. The moist western slope supports extensive and magnificent forests at middle elevations. The eastern slope supports trees more tolerant of its aridity and rockier landscape—single-leaf pinyon pine *(Pinus monophylla)*, limber pine *(Pinus flexilis)*, and Utah juniper *(Juniperus osteosperma)* to the north.

The Sierra Nevada and the Cascade Range offer an assortment of environments in stark contrast to those encountered in other regions of California. During a very brief breeding season with cold nighttime temperatures, often blistering days, and even occasional summer snows, the high-mountain species find a variety of unusual niches in which to breed: the Great Gray Owl *(Strix nebulosa)* nests in the broken trunk of a lightning-struck red fir *(Abies magnifica)*; the Calliope Hummingbird *(Stellula calliope)*, North America's smallest bird, breeds at the highest elevations; the American Dipper *(Cinclus mexicanus)* builds its nest in the splash zone of small waterfalls.

The dominant southern California mountains include the San Gabriel (highest peak 10,064 feet) and the San Bernardino

Plate 89. A Calliope Hummingbird female feeds a nestling by thrusting her bill deep into its open gape. Hummingbird males take no part in the care of their offspring.

(highest peak 11,502 ft) Mountains of the Transverse Ranges and the San Jacinto Mountains (highest peak 10,805 ft) of the Peninsular Ranges, all of which abruptly rise two vertical miles from their surrounding valleys. Sandwiched between California's arid deserts and the subtropical southern coastal hills, these isolated peaks are outliers of the southern Sierra, and share strong similarities with the Sierra's floristic and avian communities. These islands of montane habitat variably attract, or hold, small populations of montane birds such as Blue Grouse *(Dendragapus obscurus)*, Williamson's Sapsucker *(Sphyrapicus thyroideus)*, White-headed Woodpecker *(Picoides albolarvatus)*, Clark's Nutcracker *(Nucifraga columbiana)*, and Mountain Chickadee *(Poecile gambeli)*, which are far from their species' centers of distribution. Isolated populations provide interesting subjects for research into the biological processes of extinction and colonization, survivorship and recruitment of populations, effects of habitat fragmentation, and other questions known collectively as metapopulation dynamics.

We concentrate mostly on the avifauna of the Sierra Nevada,

because this range provides the most extensive core mountain habitat in California, is the best known, and is the most accessible. The Transverse and Peninsular Ranges, however, have their own unique communities and seasonal rhythms that await exploration by the curious naturalist. There are also hundreds of small ranges and peaks with their own peculiar and fascinating avifauna, many of which have only recently begun to be birded—Glass Mountain, the Greenhorns, the Marbles, to name but a few.

Zonation

The distribution of plant and animal communities along California's spine is controlled by the interacting influences of latitude, elevation, and slope. With an increase in latitude (from south to north), precipitation increases and temperature decreases. This is a gradual effect: every gain of 1,000 feet in vertical elevation is equivalent to moving 300 miles to the north. Therefore, the yellow pine (ponderosa and Jeffrey pines *[Pinus ponderosa* and *P. jeffreyi]*) forest (also referred to as the lower montane zone) occurs from 1,000 to 6,000 feet in elevation in the northern mountains, and from 4,000 to 9,000 feet in the south-

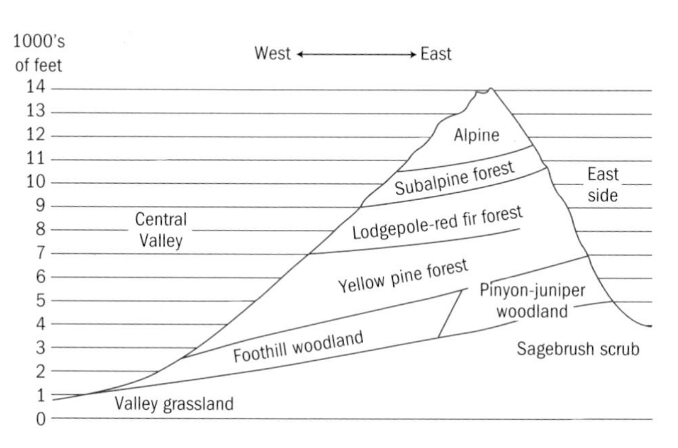

Figure 2. Biotic zonation of the central Sierra Nevada. Corresponding zones are elevated toward the south and on the east side of the Sierra Nevada.

ern Sierra. Upslope from Sacramento, ponderosa pine *(Pinus ponderosa)* is encountered at 2,000 feet elevation; to the south, approaching Sequoia National Park from Bakersfield, the transition from the foothills into the conifer zone is up at 4,500 feet. On Mount Shasta, Clark's Nutcracker ranges up into whitebark pine *(Pinus albicaulis)* at timberline at about 8,000 feet. In the southern Sierra, timberline is up at about 10,000 to 11,500 feet.

Distributional Patterns

The Sierra is at the southern limit of the great boreal forests that range northward through the Cascades and into the Canadian Rockies and Coast Ranges. It is also the southern limit of glaciers and mountain hemlock *(Tsuga mertensiana)* in North America. Where mountain hemlock ends, above Sequoia National Park, it meets foxtail pine *(Pinus balfouriana)*, the tree-line conifer of the southern Sierra. Several birds reach the southern edge of their distributional ranges here as well: Great Gray Owl, Black-backed Woodpecker *(Picoides arcticus)*, and Pine Grosbeak *(Pinicola enucleator)*. Harlequin Duck *(Histrionicus histrionicus)* also should be included in this group; however, sadly, this best-dressed of North American wildfowl has been nearly extirpated from its former breeding haunts. Though never common, it formerly nested along fast-moving streams amid the yellow pine belt (above 4,000 feet) and in Yosemite Valley proper. A few pair persisted along the upper Mokelumne River and near Yosemite through the 1970s, but the decline began as early as the 1920s (Gaines 1977). Their rarity in the southern Cascades suggests that they are absent there as well. Harlequins were never common, but formerly were found "sparingly along secluded streams," where they "seemingly share with the Ouzels a love for the noisiest parts of the river" (Grinnell et al. 1918). California's mountains can no longer provide the privacy the Harlequin needs to nest or the Eden-like natural world found by John Muir on his first visits:

> Nor was there any lack of familiar birds, bees, and butterflies. Myriads of sunny wings stirred the air. The Steller jay, garrulous and important, flitted from pine to pine; squirrels were gather-

> ing nuts; woodpeckers hammering dead limbs; water ouzels sang divinely on foam-fringed boulders among the rapids; and the robin redbreast of the orchards was in the open groves. Here no field, nor camp, nor ruinous cabin, nor hacked trees, nor down-trodden flowers, to disenchant the Godful solitude . . . no word of chaos or desolation among these mighty cliffs and domes. (Wolfe 1979)

In the Sierra Nevada, several montane birds of the western slope are largely absent on the eastern slope. Most have habitat affinities that are simply not available on the east side. Great Gray Owl requires large mountain meadows surrounded by conifer forests, a habitat present only on the gentler slopes west of the Sierran crest. Acorn Woodpecker *(Melanerpes formicivorus)*, Chestnut-backed Chickadee *(Poecile rufescens)*, and Hutton's Vireo *(Vireo huttoni)* are associated with elements of oak woodland that are interlaced with yellow pine forest. Several species prefer the moist, mixed-conifer forests that are only marginally available on the arid east side. Pileated Woodpecker *(Dryocopus pileatus)* is partial to the large lumber of giant sequoia *(Sequioiadendron giganteum)*, pine, and fir; Hammond's Flycatcher *(Empidonax hammondii)* likes shady groves, especially red fir and sequoias, but also ranges upward into the subalpine zone. Pacific-slope Flycatcher *(Empidonax difficilis)* is mostly associated with riparian thickets amid conifer forest, especially in the ponderosa pine association. The California Spotted Owl *(Strix occidentalis occidentalis)* (not to be confused with the Northern Spotted Owl *[Strix occidentalis caurina]*, a subspecies confined to the North Coast Ranges), occurs in middle-elevation forests across the western slope; this habitat is uncommon on the eastern slope, and Spotted Owls are equally rare there.

Most species confined to the eastern slope are either at the western edge of their Great Basin range, and are more attuned to the drier habitats found there, or are limited by the plant associations not available on the western slope. Some examples are Gray Flycatcher *(Empidonax wrightii)*, Pinyon Jay *(Gymnorhinus cyanocephalus)*, Black-billed Magpie *(Pica hudsonia)*, Virginia's Warbler *(Vermivora virginiae)*, and Brewer's Sparrow *(Spizella breweri)*. Another, Yellow-headed Blackbird *(Xanthocephalus xanthocephalus)*, nests almost exclusively in marshes with tall, emergent vegetation over deep water, a habitat that is common in the Great Basin but simply absent on the western slope, except at a few spots such as Sierra Valley.

Birds typical of the Cascades that follow that range from Oregon southward into California, but do not continue southward into the Sierra, are Gray Jay *(Perisoreus canadensis)*, Black-capped Chickadee *(Poecile atricapilla)*, and Juniper Titmouse *(Baeolophus griseus)*, though the latter is widespread in the adjacent Great Basin. Erratic incursions of northern birds into the southern Cascades and even the Sierra, occurring when severe weather or scarce food causes irruptions of species such as Bohemian Waxwing *(Bombycilla garrulus)*, White-winged Crossbill *(Loxia leucoptera)*, and Common Redpoll *(Carduelis flammea)*, is infrequent but highly anticipated by birders.

Some species are common in some years and absent in others. The erratic distribution of the Red Crossbill *(Loxia curvirostra)* is well known. As its name suggests, the most obvious physical characteristic of this bird is the oddly shaped bill, with the upper and lower mandibles skewed in opposite directions, seemingly deformed. This asymmetry is not a deformation, however, but ingeniously designed by evolution to allow the bird to pry open the scales of pine and fir cones and extract the seed. "In choice of nesting dates the Crossbill is probably the most erratic of all northern birds" (Dawson 1923). Having as they do a regionally variable, but locally dependable food supply, these birds nest wherever and whenever an abundant cone crop prompts them—it may be January, it may be October.

Habitats

Western Foothills (Grasslands, Foothill Pine, Chaparral)

In the foothills, oak savannah and grasslands characteristic of the Central Valley floor intergrade with foothill pine *(Pinus sabiniana)* forest and chaparral. Many of the birds that frequent this transition into the foothills are familiar to Californians. Turkey Vulture *(Cathartes aura)* and Red-tailed Hawk *(Buteo jamaicensis)* are the common soaring raptors. White-tailed Kite *(Elanus leucurus)* is uncommon, but may be seen hovering over moist fields at any season. American Kestrel *(Falco sparverius)*, Loggerhead Shrike *(Lanius ludovicianus)*, and Western Kingbird *(Tyrannus verticalis)* perch on fence posts and wires and forage over open fields; Red-shouldered Hawk *(Buteo lineatus)* and Black Phoebe *(Sayornis nigricans)* stay close to riparian corridors. In winter, Sharp-shinned Hawk *(Accipiter striatus)* and Cooper's Hawk *(Accipiter cooperii)* are as likely in the foothills as anywhere, except perhaps the Coast Ranges. In the less peopled regions, where open pasture is extensive enough for jackrabbits, Golden Eagles *(Aquila chrysaetos)* still range. Barn Owl *(Tyto alba)* is common, but seldom seen; the lower foothills may be a last stronghold outside the Imperial Valley for Burrowing Owl *(Athene cunicularia)*, but even here it is getting exceedingly scarce. Farther upslope, the wavering whistle of a Western Screech-Owl *(Megascops kennicottii)* emanates from reaching oaks silhouetted on moonlit hillsides. Where foothill pine joins the community, so does the Northern Pygmy-Owl *(Glaucidium gnoma)*. The broken forest from the foothill pine belt upslope through the mixed forest is the Pygmy-Owl's prime habitat, although the similar structured forest of the inner Coast Ranges also holds good numbers. Valley and oak woodland birds such as Yellow-billed Magpie *(Pica nuttalli)* and Western Scrub-Jay *(Aphelocoma californica)* are equally likely to be seen; yet neither easily tolerates the other, as befits the corvids.

Most of the landbirds of the lower foothills are the same as those found in the interior Coast Ranges, as both regions support similar habitat types—grassland, oak woodland savannah, and hard chaparral. The most conspicuous year-round residents include Northern Flicker *(Colaptes auratus)*, Bushtit *(Psaltriparus minimus)*, Bewick's Wren *(Thryomanes bewickii)*, Western Bluebird *(Sialia mexicana)*, American Robin *(Turdus migratorius)*, Northern Mockingbird *(Mimus polyglottos)*, California and Spotted Towhees *(Pipilo crissalis* and *P. maculatus)*, Lark Sparrow *(Chondestes grammacus)* and Song Sparrow *(Melospiza melodia)*, House Finch *(Carpodacus mexicanus)*, and Lesser and American Goldfinch *(Carduelis psaltria* and *C. tristis)*. In oak woodlands, Acorn Woodpecker and Nuttall's Woodpecker *(Picoides nuttallii)*, White-breasted Nuthatch *(Sitta carolinensis)*, Oak Titmouse *(Baeolophus inornatus)*, and Hutton's Vireo are added to the mix, and in chaparral, Wrentit *(Chamaea fasciata)*, Spotted Towhee, and Rufous-crowned Sparrow *(Aimophila ruficeps)* are found.

Plate 90. Barn Swallows build their mud-cup nests under the eaves of buildings and in cornices below bridges. These widespread summer visitors may raise several broods in a season.

Seasonally common insectivores in the wooded lower foothills include Western Wood-Pewee *(Contopus sordidulus)*, Ash-throated Flycatcher *(Myiarchus cinerascens)*, Warbling Vireo *(Vireo gilvus)*, and Orange-crowned Warbler *(Vermivora celata)*. All of the swallows except Bank Swallow *(Riparia riparia)* and Purple Martin *(Progne subis)* are common throughout the foothills in spring, summer, and early fall.

Much of the grassland of the lower foothills is dedicated to cattle grazing; Western Meadowlark *(Sturnella neglecta)* and Lark Sparrow, in particular, are associated with these grazed grasslands. Springs and stock ponds have been developed as reliable water sources for cattle, and small marshes and cattail-lined ditches have developed in association with these watering holes, inadvertently creating wetland habitat in an otherwise arid landscape. These small wetland fragments have provided habitat for some marsh birds. Most remarkably, in the mid-1990s, Jerry Tecklin, a researcher at the University of California's Sierra Foothill Research Center east of Marysville discovered small populations of California Black Rails *(Laterallus jamaicensis coturniculus)*, a species formerly thought to be restricted to tidal marshes, scattered at disparate marshes across the lower foothill region of Nevada and Yuba Counties. Some of the marshes in which rails were found were nonexistent prior to the mid-1970s; therefore, they had been colonized fairly recently. Black Rails are largely sedentary, staying put once in suitable habitat. Wetlands, however, are dynamic environments and go through cycles of desiccation, flooding, and fluctuations in food supply. Although migration and dispersal patterns of most rails are poorly understood because they fly at night, their ability to colonize newly created habitats is an example of the avian ability to adapt to changing environments. Because we have so altered the wetlands of the west, mostly destructively, such adaptability may prove the saving grace of this (and other) beleaguered species.

Middle-Elevation Forests (the "Yellow Pine Belt")

The xeric lower foothills—grasslands, blue oak *(Quercus douglasii)* woodlands, and chaparral intermixed with foothill pines—give way to a more mesic and mixed forest at middle-elevations that may span 4,000 vertical feet, depending on latitude. In drier microclimates, as on south-facing slopes, the deciduous black oak *(Q. kelloggii)*, ponderosa pine, incense-cedar *(Calocedrus decurrens)*, and sugar pines *(Pinus lambertiana)* are interspersed and moderately spaced, with a generous shrubby undergrowth. In damper conditions, such as in canyon bottoms and on north-facing slopes, where evapotranspiration rates are lower, the trees are more tightly packed; Douglas-fir *(Pseudotsuga menziesii)*, bigleaf maple *(Acer macrophyllum)*, and California bay *(Umbellularia californica)* join the plant community, providing more canopy cover and stratification of the forest. These middle-elevation forests, for all their diversity, mark the transition from lowlands, of which hardwood oaks are the signature trees, to a montane environment, where conifers dominate. This transition is mirrored in the small owls of the genus *Otus* as well. Western Screech-Owl, a bird primarily of oak woodlands, reaches its upper limit in the yellow pine belt, where Flammulated Owl *(Otus flammeolus)* becomes common.

The diminutive Flammulated Owl (about 5 percent the body weight of a Great Horned Owl *[Bubo virginianus]*) is "perhaps the most common raptor of the montane pine forests of the western United States" (McCallum 1994). Also, it is probably the least often encountered and its habits the least well known. The Flam is strictly insectivorous, eating mostly moths and various arthropods. It is probably migratory, a long-distance, north-south migrant, although it has been speculated that it goes torpid. It has also been suggested that the bird is semicolonial, based on observations of clusters of birds interspersed among unoccupied habitat, but this question has not been resolved. It nests in the Sierras and Cascades as well as in the interior Coast Ranges, the Great Basin mountains, and the Transverse and Peninsular Ranges; throughout, it is closely associated with yellow pine in "dry, open ponderosa pine or forest with similar features" (Winter 1974).

Although slopes with southern exposure can be harsh and hot, the varied topography of middle-elevation forests also offers cool, shady groves that accommodate the thermoregulatory needs of Spotted Owls. The California Spotted Owl resides in mixed conifer forest of the Sierra Nevada from the Pit River (Lassen area) south through the Greenhorn Mountains at the southern end of the Sierra, with a few outlier sites east of the Sierran crest in pockets of mixed conifer forest. The distribution is disjunct, as other populations are distributed in the South Coast Ranges as well as in the Transverse and Peninsular Ranges.

More than three-quarters of Spotted Owl territories in the Sierra Nevada are in mixed conifer forest, with fewer numbers downslope in ponderosa pine and hardwood forest or upslope in red fir. At higher elevations, flying squirrels are the preferred prey; at lower elevations, woodrats become more important. The abundance and availability of prey determines the owl's nesting success, and the abundance of prey is determined by the extent and diversity (edge ratio) of the habitat. Nesting trees tend to be large and located in decadent forests, with many dead and decomposing trees. The forest canopy needs to be nearly closed, to provide protection from severe weather and predators. Large amounts of suitable habitat around the nest site promotes greater nesting success and, therefore, sustainable populations of owls.

These mixed, middle-elevation forests, with their diversity of commercially valuable tree species, are also heavily logged. Current logging practices disrupt Spotted Owl habitat, and, therefore, owls are at risk, especially on private and managed forest lands. Demographic studies indicate that Spotted Owls are declining at an alarming rate, approaching 10 percent per year in some areas. Numbers are most at risk in the northern Sierra na-

Plate 91. A White-headed Woodpecker emerges from its nest hole in the trunk of a fallen giant sequoia.

tional forests (managed forest) lands; the region of Sequoia National Park seems to be the most stable.

The combination of diverse vegetation, relatively equitable climate, and moderate precipitation provides a hospitable environment for many mountain species. The transition from hardwood (oaks) to softwood (conifers) in middle-elevation forests accommodates the boring needs of Pileated Woodpecker and White-headed Woodpecker, each of which is resident year-round, but may move up- or downslope as weather conditions change seasonally. In the summer months, small flycatchers partition out the microclimates, with Dusky Flycatchers frequenting the drier brushy slopes and Pacific-slope Flycatchers preferring the damper, shady forest. Larger flycatchers separate out vertically, with Western Wood-Pewee usually foraging at middle elevations within the forest canopy or at openings on the edge, and Olive-sided Flycatcher *(Contopus cooperi)* sallying out from the uppermost branches of snags and conifers.

Where conifers dominate, Violet-green Swallow *(Tachycineta thalassina)* becomes more common than Tree Swallow *(T. bicolor);* Steller's Jay *(Cyanocitta stelleri)* largely replaces Western Scrub-Jay; and Mountain Chickadee, Red-breasted Nuthatch *(Sitta canadensis),* Brown Creeper *(Certhia americana),* and

Golden-crowned Kinglet *(Regulus satrapa)* join the community. Creepers are regularly found in oak woodland in winter, but prefer conifers for breeding. As in the Coast Ranges, Winter Wrens *(Troglodytes troglodytes)* skulk among ferns and riverbank tangles in the shaded understory. Like the flycatchers, the insect-eating vireos and warblers also separate into available niches. The Cassin's Vireo *(Vireo cassinii)* inhabits the forest where oaks and conifers intermix, whereas the Warbling Vireo is most common in deciduous broad-leaved trees—alders, maples, and aspens. Among the warblers, both the Black-throated Gray Warbler *(Dendroica nigrescens)* and the Nashville Warbler *(Vermivora ruficapilla)* are in the drier stands of forest or at the forest edge; the former more associated with oaks, the latter more prone to favor open forest amid the shrubby understory of manzanita, snowberry, and the like. Hermit Warbler *(D. occidentalis)* forages and nests higher up, in the upper limbs of firs and pines, in company with the Golden-crowned Kinglet. Audubon's Warbler *(D. coronata audubonii)* takes on this role at higher elevations (especially in lodgepole pine *[Pinus contorta* subsp. *murrayana]* forests), whereas Hermit Warbler prefers the diversity of vegetation at middle elevations. Although each species finds its own peculiar niche, each also overlaps with the other, and the territory of the Black-throated Gray Warbler may dovetail subtly with that of the Hermit Warbler and the Yellow-rumped Warbler *(D. coronata)* where the diversity of habitat and the abundance of resources allows.

Western Tanager *(Piranga ludoviciana),* arguably the most stunning of American passerines, reaches its peak abundance in the mixed conifers, where it keeps to the middle and higher branches of the trees in fairly open woodland.

> The deep blues of the Steller's Jays, the amber hues of Evening Grosbeaks and the rosy reds of Cassin's Finches, Pine Grosbeaks and Crossbills are resplendent, of course, but they are both of northern skies and northern sunsets. Audubon's Warblers, Hermit Warblers, and especially male Western Tanagers add a touch of tropical color and splendor to the summer coniferous forests. (Gaines 1977)

Middle-elevation forests produce quantities of seeds and berries and support birds with stout bills that are built for harvesting these rich sources of protein and sugar. Black-headed

LATE AUGUST. LEAVING THE MOUNTAINS, RIM OF THE WORLD.

Descending the west slope of the Sierra, we pull off the shoulder of the road to stretch and get a last breath of that pure, invigorating mountain air before dropping below the snow zone into the thicker atmosphere of Central Valley. A recent burn has left leafless skeletons of manzanita and scrub oak crisp and contorted against a soft green backdrop of the straight-spined Douglas-fir that the fire has passed by. I begin whistling with the cadence of dripping water, slow and monotonous, trying to mimic the call of the Northern Pygmy-Owl, hoping to lure one in. It looks like good habitat. Almost immediately, we hear the nervous twittering of Bushtits and then the fast-paced sputtering of Pygmy Nuthatches. Then, we hear a single whistle, almost identical to mine, but slightly harsher and hollower, coming from the fir forest. "Ah, Pygmy-Owl," I blurt out, before realizing I've been fooled again . . . it's a Steller's Jay mimicking my imitation, but more convincingly. Soon other birds are flitting through the branches–Bushtit, vireos, warblers, sparrows, and one sassy wren. All have appeared at the forest edge, scolding the interlopers, alerting each other to our presence.

We turn to leave, and only then notice a small bird, similar in size and shape to a pine cone, perched on a branch directly overhead. It's a Pygmy-Owl, peering down on us with almost fierce, unblinking eyes that seem to say–"You talking to me?"

Grosbeak *(Pheucticus melanocephalus)* occurs in relatively low-elevation oak and mixed-hardwood forest. Purple Finch *(Carpodacus purpureus)* tends toward the shadier groves of mixed-conifer forest, Chipping Sparrow *(Spizella passerina)* and Dark-eyed Junco *(Junco hyemalis)* range across several altitudinal belts, and both prefer open forest floor and dry meadows. The grosbeaks are fairly common in summer, and their melodious and ebullient song announces the height of the breeding season. Finches are variably present—abundant some years, less so others—and juncos are one of the most frequently encountered Sierran birds, especially in winter. Chipping Sparrows were once as abundant as juncos; however, they have become increasingly scarce for reasons that are not entirely clear.

Red Fir Belt

Above the drier, more open yellow pine belt, beginning at about the 7,000 foot contour in the Sierra, somewhat lower in the Cascades and Klamath Mountains, precipitation increases, snow lasts longer, and the soils become richer and more loamy. Here, lush conifer forests take on their full grandeur. Red fir, with its deep reddish brown and deeply furrowed bark, adorned with luxurious hanging tufts of bright green wolf lichen *(Letharia vulpina)*, is the most characteristic tree. Sugar pines and giant sequoia also occur here. This zone is restricted mostly to the damper western slopes. This damp forest is in the maximum snow zone, hence the pyramidal structure of the trees and the near absence of broad-leaved hardwoods, whose reaching limbs could not support the heavy snow load. "When one enters these woods, most of the birds of the lowland, Mediterranean climate . . . are left behind at lower elevations" (Gaines 1977).

As mentioned above, several northern species whose ranges extend northward through boreal forests into Canada reach their southern distribution in the Sierra. One such species is the Blue Grouse, which is most closely associated with the red fir forests, although it ranges farther upslope as well. Three subspecies occur

Plate 92. The Blue Grouse is a secretive inhabitant of mature mountain conifer forests, often revealing its presence only by a low, hooting call.

in California: Oregon Blue Grouse *(Dendragapus obscurus fuliginosus)* in the northwest Coast Ranges; Sierra Blue Grouse *(D. o. sierrae)* in the Siskiyous, Cascades, Warner Mountains, and northern Sierra; and the Mount Pinos Blue Grouse *(D. o. howardi)* in the southern Sierra and montane islands of the Tehachapi Mountains and Mount Pinos, although these relict populations may be nearly extirpated. The following refers to the Sierra race.

The grouse apparently overwinters in the subalpine forests, the low-hanging boughs of the conifers providing some shelter from the winter storms. It may subsist on fir and pine needles and even lichen during the winter, but in summer and fall it forages on the ground "around the margins of open spots and rocky moraines" (J. Muir, in Grinnell et al. 1918) on thimbleberries, manzanita berries, seeds, and insects. The hooting of the male in the spring and early summer is a unique mountain sound, "in quality . . . likened to beating on a sodden wooden tub," usually broadcast from a tree top, but ventriloquial and hard to locate. Following the breeding season the males migrate upslope to the timberline, to be followed later in the season by the females and their broods after they have visited the alder and willow thickets (Grinnell et al. 1918).

The scolding of the Steller's Jay often signals the presence of a Northern Goshawk *(Accipiter gentilis),* another year-round resident in these dense forests; the goshawk's primary prey is the grouse, but undoubtedly it finds chipmunks and chickarees as well. Midwinter, when snow is deep, goshawks may move downslope or even to the lowlands in search of prey more easily caught. In the late summer, when the Blue Grouse take their broods to the tree line and barren moraines, it is "possible to find a goshawk on an inconspicuous perch . . . waiting to take advantage of the slightest let-down in the vigilance of the feeding covey" (Gabrielson and Jewitt 1940).

With the exception of the Downy Woodpecker *(Picoides pubescens),* which rarely ranges above the low-elevation riparian forests, most of the woodpeckers that occur in the lower elevation mixed forest also range upward into the red fir–sequoia belt. Red-breasted Sapsucker *(Sphyrapicus ruber)* tends to be associated with riparian habitats—alder, willow, cottonwood—as well as conifers, especially incense-cedar at forest edges and clearings.

Contrary to the typical woodpecker image, White-headed Woodpecker forages primarily on living trees, from the yellow

SAPSUCKER HYBRIDIZATION

The Red-breasted and Red-naped Sapsuckers *(Sphyrapicus nuchalis)* were formerly considered subspecies of the Yellow-bellied Sapsucker *(S. varius)*, which inspired a bumper sticker popular among birders, but obscure to everyone else: "Split the Sapsuckers." Red-breasted, a bird broadly distributed through the Sierra Nevada and North Coast Ranges, was split off as a separate species in the 1983 American Ornithologists' Union Check-list of North American Birds. Subsequent study (1983 to 1985) determined that Red-naped Sapsucker, which occurs in scattered populations in Great Basin mountains, was also a separate species, more closely related to the Red-breasted than the Yellow-bellied Sapsucker. The three-species sapsucker complex is thought to constitute a superspecies, with limited interbreeding among them. In California, Red-naped and Red-breasted Sapsuckers come into contact in the Warner Mountains and rarely on the east slope of the Sierra, where hybridization has been documented. Hybridization is relatively common in the class Aves, with perhaps as many as 10 percent of North American species producing hybrids. Williamson's Sapsucker *(S. thyroideus)* and Red-breasted Sapsucker have also hybridized, though less commonly.

pine belt up into the red fir forest. It is particularly associated with giant sequoia groves, where it often nests low in the trunk. It forages mostly on bark insects, but also apparently eats large numbers of pine seeds, at least in those portions of its range in California.

Mountain Chickadee, Golden-crowned Kinglet, Red-breasted Nuthatch, and Brown Creeper are year-round residents, and the damp conifer forest is the business district for each species. The chickadee is the hardiest of the group, staying within the conifers through winter, flocking together to find scarce food, a task at which it must excel because it is the commonest bird in the midwinter snow zone. In the fall, chickadees begin to cache conifer seeds for use later in the season when finding food becomes more difficult. "The need to defend dispersed seeds promotes group territoriality and hierarchical social organization" (McCallum et al. 1999) so, although chickadees nest as independent pairs, after breeding they gather in flocks for the winter.

Plate 93. A Brown Creeper emerges from its nest located in a tree niche. Creepers are found from coastal woodland to mountain forests, scuttling over tree bark in search of spiders and insects.

Plate 94. The evocative call of a Mountain Chickadee is a feature of montane coniferous forest. Chickadees nest in tree cavities, often an abandoned woodpecker hole.

Plate 95. By late spring, Hermit Thrushes are delivering food to young in nests hidden in mountain canyon thickets.

But their diet is not limited to seeds. Chickadees are opportunistic feeders, variously taking moth and sawfly larvae, beetles, aphids, scale insects, and spiders, as well as seeds. Chickadees also range into lodgepole and pinyon–juniper forests on the eastern side. The kinglet, the nuthatch, and the creeper may also brave the Sierran winter, but in smaller numbers; most move downslope to the lowlands. All four are common nesters in the mixed conifer, sequoia, and red fir forests, but the kinglet is the most numerous. Like the Hermit Warbler, however, kinglets stay in the highest branches, busy amidst the lichen forests.

A sound that reaches its crescendo in the deep conifer forest is that of another mountain recluse, the Hermit Thrush. The forest thrushes (genus *Catharus*) are known for the quality of their voices, and the Hermit Thrush has one of the richest of the tribe, often described as flutelike, with the first notes of each series held for a moment, followed by rising and falling tones that reminded early ornithologist Leon Dawson of a requiem: "The song of the Hermit Thrush is a thing apart. It is sacred music, not secular . . . No voice of solemn-pealing organ or cathedral choir at vespers ever hymns the parting day more fittingly than this appointed chorister of the eternal hills" (Dawson 1923).

One of the rarest, and certainly the most impressive bird of the California mountains is the Great Gray Owl. Subsequent to its discovery in the Sierra in 1914, the Great Gray was found to be a permanent, though rare, resident, associated exclusively with moist meadow systems in and around Yosemite National Park. Less than 30 pairs are thought to occur in the state, and the population is listed as endangered.

Great Gray Owl favors forest openings, especially montane meadows surrounded by pine and fir forest. Its movements and breeding success are determined by the availability of its rodent prey; in heavy snow years, it may move downslope, where the snow cover is thinner and rodents are more easily caught. In low rodent years, the owls may forego breeding altogether. Great Grays tend to hunt from low perches, watching or listening downward intently. Pocket gophers and meadow voles compose more than 80 percent of the diet in California (Winter 1986, in Bull and Duncan 1993). Although the Great Gray seems larger than Great Horned Owl, it is at least 15 percent smaller; its abundant feathering, however, makes it appear larger than it is. This thick feathering also allows it to survive the colder habitats it favors. But this adaptive advantage may become a liability at the southern edge of its range in California, where overheating may be a danger in summer. Like the Spotted Owl, Great Grays adjust to varying temperatures behaviorally, choosing roost sites to exploit microclimates. Great Grays often nest atop broken off trunks, and the nestlings have "wonderfully cryptic gray and white coloration which camouflages them well in the forest environment where large shadows play over bark, lichen, and moldy tree trunks" (Bull and Duncan 1993).

> The discovery of the Great Gray Owl in the Yosemite section was one of the most notable events of our field experience . . . the bird was apparently quite at home, and nesting. No previous record of the breeding of this northern species of owl south of Canada is known to us. . . . On the day of discovery, however, a diminutive kinglet pointed the way, and really deserves all the credit. (Quoted in Dawson 1923)

Townsend's Solitaire *(Myadestes townsendi)*, "this poet of the solitudes" (Dawson 1923), nests in open areas, especially amid

Plate 96. A Townsend's Solitaire nestling begs for food in its well-hidden ground nest. Solitaires are common in juniper forests when the berries to which they are partial are available.

junipers growing along glacier-polished canyon slopes, or in forests of mixed conifer and red fir interrupted by rocky outcrops, burns, or natural openings in the forest. Berry-producing shrubs are usually present—manzanita, serviceberry, or snowberry. Such open habitat often occurs at the upper edge of the mixed conifer forest, before the transition into denser lodgepole pine stands. The Solitaire seems an amalgam of other birds—a cross between a bluebird, mockingbird, and flycatcher—as much because of its behavior as its appearance. "In the rocky solitudes of the Garden of the Gods it is said that the solitaire's voice is sometimes all that breaks the silence" (Bailey 1902). It is, in fact, a thrush, most closely related to the bluebird.

Lodgepole Pine Forest

A distinctive feature of the High Sierra timberline is the lodgepole pine, a tree well adapted to the damp environment of alpine meadows and lake basins and the most common tree just below timberline in the subalpine zone. Lodgepoles extend downslope, along streams and in frost pockets, to the yellow pine belt. Black-backed Woodpecker may be the bird most associated with lodge-

Plate 97. To feed its nestlings, a Black-backed Woodpecker collects large insect larvae, which it carries lengthwise in its bill.

pole pine forest on the high western slopes of the Sierras and the Cascade Range. It is also present in the Siskiyou and Warner Mountains, where it is associated with red fir and yellow pine snags, especially those in early stages of decay. Black-backs are a disturbance-adapted species, attracted to insect infestations, forest burns, and blowdowns. In fact, White-headed Woodpecker and Hairy Woodpecker *(Picoides pubescens)* are often found along with Black-backs in areas so impacted. A fire in the Jeffrey pine forest on the eastern slope near Mono Lake in April of 2001 attracted both Black-backed and White-headed Woodpeckers, species generally rare or absent on the east side (Dunn 2002).

The sapsuckers are zoned according to elevation and habitat, with Red-breasted preferring the aspen and lower pine forests and Williamson's Sapsucker the higher lodgepole zone. Like the Red-breasted Sapsucker, Williamson's relies on living trees, in which it bores rows of holes, uniformly circling the trunk or limb. The sapsuckers then return periodically to these "tattoos" to feed on the flowing sap. In addition to lodgepoles, Williamson's uses western white pine *(Pinus monticola)* and mountain hemlock; on the eastern slope it tends to frequent Jeffrey pine. In the cold winter months sap moves slowly (or not at all), but some early references suggest Williamson's remained year-round at these high elevations. I've seen this species in mixed conifer forest in Yosemite Valley in midwinter foraging on incense-cedar, indicating post-breeding movement to lower elevations. Within the woodpecker family, the male usually has some (or more) red in his plumage, but Williamson's Sapsucker is the most sexually dimorphic of all the North American Picidae. So different are the plumages that early ornithologists classified the male and female as different species! Usually nature gives clues to the reason for such uniqueness, but the reason for this distinction in this species remains mysterious.

One of the common insects in this forest is the lodgepole needle miner *(Coleotechnites milleri)*, which has a two-year life cycle: one year larval, one adult. Mountain Chickadee is attuned to the this cycle and concentrates on the resource in alternate years, deftly peeling back pine needles to extract the needle miner larvae (McCallum et al. 1999). One cannot help but wonder what other rhythms, as yet undiscovered, make up the seasonal pulse of the mountain forest. To survive and thrive in such an environment, each species must have its own natural rhythms, finely orchestrated and unique to itself.

Pine Grosbeak is a resident in the central Sierra (absent from the southern Cascades and the southern California mountains), where it nests in the 7,000- to 10,000-foot range. The California race *(Pinicola enucleator californicus)* is segregated from its more northern relatives, and is endemic to the state. Within its restricted range, however, it wanders widely in search of an abundant cone crop. This seedeater's strategy is akin to that of Red Crossbill, Evening Grosbeak *(Coccothraustes vespertinus)*, Purple Finch, and Pine Siskin *(Carduelis pinus)*; however, those species wander much farther afield from their core breeding areas. Pine Grosbeak occurs year-round in moist portions of open forest of red fir and lodgepole pine. It forages mostly low in foliage or on the ground, where it uses its stout bill to nip the buds and growing tips of conifer branches and to crush seeds. Though rarely encountered, this bird is surprisingly calm when approached in the forest where it usually forages in pairs, or, in winter, in small groups of a dozen or so birds. Pine Grosbeaks generally avoid disturbed areas, especially logged-over forest tracts. Because there are few lower elevation records in winter, we assume they spend the winter in the fir and pine forests, but as David Gaines (1977) wondered, "The winter whereabouts of Pine Grosbeaks is one of the Sierra's intriguing unsolved mysteries. . . . They must be up there . . . somewhere!" Another secret known only to these grosbeaks is why they occur in the Sierras but not in the Cascades or the Warner Mountains.

Red Crossbill, perhaps the ultimate conifer cone specialist, also stays at higher elevations, where lodgepole pine seeds are abundant. On the western slope of the mountains it rarely drops below the red fir belt; on the eastern slope it visits the pinyon–juniper and yellow pine forest. The variability of cone production, caused by variation in weather cycles, requires erratic behavior by birds that rely on the seeds produced by the cones. To find the variably abundant crops, crossbills travel in large flocks over long distances, cavalcading through the treetops, chattering among themselves ('kip-kip, kip-kip"), on constant search. They also may nest at any time of year, in any place where cones may be available.

Like the crossbill, Pine Siskin is most common in the lodgepole pine belt, is highly gregarious, wanders widely, shifts nesting locality from year to year, and swings wildly in abundance from year to year. Unlike the crossbill, the siskin has a fairly broad diet, varying among pine, alder, and thistle seeds. Perhaps as a result of this dietary flexibility, the siskin is as common on the eastern as on the western slope.

Cassin's Finch *(Carpodacus cassinii)* finds its most hospitable environment in the lodgepole belt throughout its range in the mountains of the west: "In choice of habitat Cassin's fall between the Purple Finch, which dwells in moist well-shaded forests at lower elevations on the west slope, and the House Finch which dwells in arid basin scrub east of the Sierran escarpment" (Gaines 1977). In the Sierra and Cascades, Cassin's distribution is similar to Pine Siskin, from the lodgepole belt on the western and throughout the forested zones on the eastern slope; however, unlike siskins and crossbills, Cassin's Finch does not seem to wander as widely or swing as erratically in numbers. Perhaps its reliance not only on seeds but also on buds and berries affords its populations some degree of stability.

Subalpine Meadows and Slopes

Above the lodgepole forests, whitebark pine *(Pinus albicaulis)* and splendidly deformed mountain junipers define the timberline.

These sparsely treed granite or volcanic slopes in any of California's highest mountains are the domain of the Clark's Nutcracker, an outcast relative of the crow and raven. Like his cousins, the nutcracker is a hardy and noisy bird, but there the similarity ends. In behavior, the nutcracker is more akin to the Acorn Woodpecker, but rather than acorns, the nutcracker hordes pine seeds. Like the woodpecker, nutcrackers also catch flies from prominent perches on warm days to augment their seedy diet. The nutcracker's primary provider is the whitebark pine, the characteristic tree of the subalpine zone. This is the multiple-limbed, storm-sculpted, contorted tree of the highest mountains, found only above 6,000 feet in the Cascades and upward to 12,000 feet in the southern Sierra. More than 90 percent of the whitebark's seeds are cached, mostly by nutcrackers, which carry mouthfuls (stored in a pouch under the tongue) as far as eight miles to their personal, secret stash site. These storage sites are dug in the ground or trees, usually at lower elevations, so that they may be retrieved in winter when food is scarce. Although a common bird of the western slope, nutcrackers typically cross the divide to cache their seeds and breed on the eastern slope of the Sierra. They are most common in the pinyon and Jeffrey pine belt on the east side in winter, presumably because the lighter snowfall allows more access to their cache of seeds. Nutcrackers lay their eggs earlier than most Sierran birds, fledging young in April, and therefore roving family groups are conspicuous in spring.

The role of both Clark's Nutcracker and Pinyon Jay as planters of pine seeds is worthy of note. Both are members of the corvid family, and like their relative the Western Scrub-Jay, who buries acorns in the Coast Ranges and foothills, each species is prone to caching seeds in the ground and relocating them by memory. Although the birds return to feed on their seed caches, large numbers are never retrieved, and some undoubtedly germinate the following summer. Through this behavior, the birds and their forage tree are allied in a mutually beneficial relationship: the birds provide a dispersal mechanism for the trees, and the trees provide an abundant food source for the birds.

In summer, below the melting glaciers or where water seeps out of talus slopes, moist meadows and seepage slopes develop. Here, dwarf alpine or Sierra willow *(Salix eastwoodiae)* forests, intermixed with bog laurel *(Kalmia polifolia)* and red mountain heath *(Phyllodoce breweri)* attract breeding warblers and sparrows and provide cover for birds that drift upslope following the breeding season. Orange-crowned Warbler, Nashville Warbler, Wilson's Warbler *(Wilsonia pusilla)* and MacGillivray's Warbler *(Oporornis tolmiei)*, along with Chipping Sparrow and Mountain White-crowned Sparrow *(Zonotrichia leucophrys oriantha)* find their way here. These moist meadows also are bedazzled with an array of wildflowers—paintbrush, penstemon, scarlet gilia—which attract migrant hummingbirds in midsummer.

Alpine Zone

At the upper edge of the tree line on the western slope of the Sierra, only scattered trees—whitebark pine in the north, foxtail pine farther south—give way to nearly barren granite scree. Limber pine *(Pinus flexilis)*, which grows like a large sprawling shrub, is rather common east of the crest. Stunted and deformed trees give way to alpine grasses and stony fells above the tree line. Few birds venture into this rarefied atmosphere, where wind and temperature can be extreme. But around the north-facing cirques, at the bases of vertical escarpments where only the sparsest grass or lichen can take hold, or at the edge of a melting snowfield or gla-

Plate 98. Gray-crowned Rosy-Finches are birds of the high mountain slopes and associated snow patches. In winter, when their high-altitude feeding grounds are snow-covered, flocks can occasionally be seen at lower elevations where the snow cover is lighter.

cier, the Gray-crowned Rosy-Finch *(Leucosticte tephrocotis)* has found its niche. Here, it plies the barren ground for frozen seeds and insect shards, or picks at ice worms in the softened snow. The winter haunts of the rosy-finch are poorly known, but enough midwinter sightings at high mountain passes suggest that flocks tend to stay above tree line even in winter months. Other sightings indicate that when they move downslope, it is most often east of the crest.

The East Side

The forests of the eastern Sierran escarpment are an intermixture of Great Basin flora and drought-tolerant Sierran conifers. Though sometimes diverse in species composition, these forests are patchy and relatively open. Through the lower elevations, yellow pine and single-leaf pinyon dominate, and with increasing elevation, a potpourri of pines—Jeffrey, lodgepole, limber, western white, whitebark, and foxtail—occur in varying combinations depending on soil conditions, microclimate, and competition.

Pinyon-Juniper-Sagebrush Zone

The characteristic bird of these eastern slope forests in the aptly named Pinyon Jay, a corvid as partial to pinyon pine seeds as its cousin, Clark's Nutcracker, is partial to whitebark pine seeds. Although the pinyon is a small tree with small cones, its seeds are large and produced in profusion. So abundant are pinyon seeds that they provided an important food staple for Native Californians. Cone production of conifers is highly variable from year to year, and Pinyon Jays deal with this unpredictability by wandering widely in search of sufficient mast to satisfy their appetites. Nomadic flocks can be quite large, involving 100 birds or more. When Pinyon Jays find a foraging site, however, they tend to spread out through the forest, several birds to a tree, staying in constant communication with their distinctive calling—"queh, queh, queh." When feeding on the ground, which they frequently do, the flock may be more tightly packed, moving along in a leapfrogging fashion, "the members at the rear flying over and past those ahead of them on the ground" (Miller and Stebbins 1964).

Single-leaf pinyon, most abundant in the southern Sierra and Transverse Ranges, grows from as low as 3,000 feet elevation upward to 10,000 feet, so Pinyon Jays are encountered mostly within that vertical zone. Flocks occasionally wander outside this range, probably when seed crops are scarce.

The eastern slope is in the rain shadow of the Sierras and the Cascade Range, and therefore is a much more arid environment than the western slope. Because the eastern slope is drier, the forest is generally more open and the understory less dense. As high upslope as the lower edge of the tree line, the pinyon–juniper forest is interlaced with sagebrush, so montane bird communities overlap with those of the Great Basin. Therefore, common Great Basin birds such as Gray Flycatcher, Green-tailed Towhee *(Pipilo chlorurus)*, and Brewer's Sparrow extend upward onto the eastern slope to high elevations (about 9,000 feet) in summer. The yellow pine forest (Jeffrey and ponderosa) also intermixes with the pinyon–juniper and sagebrush, and these ecotones support their own special complement of species. After the breeding season, there is an upslope drift on the eastern slope as on the western slope, and these species occasionally, or perhaps regularly, cross over the passes into subalpine and high-altitude willow thickets. It should be noted that in the southern Sierra, as on the Kern Plateau, these generalizations break down. There, the combination of milder climate, less-precipitous slopes, and more communication among habitat types allows species such as Pinyon Jay and Brewer's Sparrow to cross over onto the western slope commonly.

Where chaparral occurs either uniformly, or as an understory to widely spaced forest, Sage Thrasher *(Oreoscoptes montanus)*, Green-tailed and Spotted Towhees find shelter. Surprisingly, House Finch, a common breeding species at lower elevations on the eastern slope, is largely absent from western slope chaparral. Common Poorwill *(Phalaenoptilus nuttallii)* is a summer resident of the shrubby slopes, more common to the east than to the west.

Riparian and moist meadow associations on the east side concentrate several species absent, or rare, west of the crest: Black-billed Magpie, Swainson's Thrush *(Catharus ustulatus)*, Western Meadowlark, Yellow-headed Blackbird, Lazuli Bunting *(Passerina amoena)*, and Savannah Sparrow *(Passerculus sandwichensis)*. Calliope Hummingbird is most common in riparian areas

Plate 99. The Green-tailed Towhee inhabits mountain chaparral, where it builds a cup-shaped nest on or near the ground.

on the eastern slope, less so in the ponderosa pine to red fir zone of the western slope. Common Nighthawks *(Chordeiles minor)* often course over most of the open eastern slope valleys in summer, but are rare west of the crest.

At the lower edge of the eastern slope forests, where the Sierra meets the Great Basin sagebrush steppe, "Species richness of mammals . . . is among the highest in California and in North America" (Shuford and Fitton 1998). One raptor who exploits this rodent abundance most effectively is the rather elusive Long-eared Owl *(Asio otus)*. Interestingly, birders, most of whom live in the lowlands, tend to think of Long-eared Owls as "irruptive," their numbers and occurrence fluctuating with rodent populations. This appears to be the case in the lowlands, where the preponderance of rodents are short-lived microtines (especially meadow voles), a group prone to boom-and-bust cycles. But in the mountains, the rodent population is dominated by long-lived species, such as pocket mice (*Perognathus* spp.) and kangaroo rats (*Dipodomys* spp.). This group of *heteromyid* rodents tends to support much more stable populations than the lowland microtines. This prey dependability, in turn, supports a more stable population of predators. Indeed, a study of raptors on Glass

Mountain in the Mono Basin found that the center of abundance of Long-eared Owl in California is this Great Basin–Sierran ecotone (Shuford and Fitton 1998).

Watercourses

From their origins in ice fields and glacial meltwater, through alpine fells and meadows, downward through rills and ravines, over waterfalls, into cascades, rapids, and meandering streams, the Sierran watercourses traverse all mountain habitats. Certain birds are particularly associated with these waterways, whatever their elevational affinities, from the headwaters to the lowlands. As discussed above, Gray-crowned Rosy-Finch habituates the headwaters of the Sierran drainages.

The ultimate mountain waterbird is in fact a passerine. The American Dipper, once known as the "Water Ouzel," has adapted its behavior to a watery life, not only dipping, but also swimming, diving, and even walking along the bottom of the stream in search of food. The dipper has become one with the mountain stream, spending its entire life along the bank and foraging on its

Plate 100. American Dippers usually build their nests on rock ledges behind waterfalls, and sometimes on the underside of a bridge over a mountain stream.

rocks and boulders, both above and below the surface, for its insect prey—caddisfly and stonefly larvae, trout food. It moves downstream only when the surface freezes over, and then only as far as it needs to find running water. It even builds its nest within the splash zone of small cascades and waterfalls, attached to the side of a boulder or undercut bank where the mossy covering will stay damp, moist, and green.

> The life of an ouzel seems an echo of the mountain streams . . . His song is low, sweet, and fluty, expressing the very heart-peace of nature, steadfast as a star in its shining, though dwelling in the midst of the blare and glare of wild torrents . . . a dusky, dainty bird that sings deliciously all winter. No matter how frosty or stormy, go to the riverside and you will hear him . . . whole-souled enthusiastic music that, despite the cutting wind about your ears, will make you fancy you are in a flowery garden . . . This one bird makes a summer any time of year. (John Muir in Wolfe 1979)

Another bird closely associated with mountain streams is the Spotted Sandpiper *(Actitis macularia),* which nests along the shore of slow-moving streams or placid mountain lakes from the conifer zone up to tree line. It is seen most often bobbing in place at the water's edge, or flying along the river with shallow wing beats in a characteristic and easily recognized flap-and-glide motion. Another shorebird, the Killdeer *(Charadrius vociferus),* also nests on gravelbars or barren ground along the stream, but it is much more general in its nesting requirements, more widely distributed, and more habituated to human activity, sometimes even nesting in parking lots or pull-outs.

One of the most elusive of California birds, the Black Swift *(Cypseloides niger),* also builds a mossy nest behind waterfalls, but only the most dramatic and raging torrents, such as those that spill over the sheer cliffs of Yosemite. This largest North American swift is also the swiftest and the most aerobatic, soaring so high and so fast that it is seldom seen by mere mortals. Early accounts refer to the "Black Cloud Swift," a wonderful reference to its lofty habitat, high above the skies plied by other aerial insectivores. Most birders only glimpse this aerial acrobat as it darts behind a waterfall at Lassen or Yosemite, or at one of the few other traditional mountain nesting sites. Swifts have been reported as having the remarkable habit of mating on the wing.

Plate 101. A Spotted Sandpiper usually lays its eggs in a depression near the shore of a mountain lake or stream. Like some other species that frequent water edges, this sandpiper has the habit of continually bobbing.

Whether these midair antics are copulation or aggressive interactions does not lessen the flourish of Dawson's (1923) description: "Now a sable lover made soft proposal to a dusky mate, the while they scudded across the heavens like twin meteors."

The Willow Flycatcher *(Empidonax traillii)* is "partial to the shrubby willows that line the languid streams and meadows" (Gaines 1977), and so qualifies as a streamside species. This bird, another of the difficult to identify *Empidonax* flycatchers, is one of the small neotropical migrants that has suffered the dual impacts of reduction in willow cover because of clearing, grazing, and damming of rivers and the concomitant increase in the Brown-headed Cowbird *(Molothrus ater)* (discussed in "Birds of the Central Valley and Delta"). The Willow Flycatcher is a favored host of this brood parasite, and the host has suffered accordingly. This flycatcher is now excluded from its former haunts in the lower foothills, and is increasingly rare in the higher mountains. Its listing by the state as endangered has increased attention to its habitat and nesting requirements, at least on public lands, and there is still hope that remedial action, such as restricting grazing from streamside corridors on public lands, will restore critical habitat values and reduce its susceptibility to cowbird parasitism.

Plate 102. Willow Flycatchers build relatively unconcealed nests in the forks of forest saplings and are especially vulnerable to cowbirds.

Plate 103. A Belted Kingfisher may occasionally feed its young on aquatic animals other than fish. Here the prey is a larval Pacific Giant Salamander, *Dicamptodon ensatus*.

At lower elevations Belted Kingfisher *(Ceryle alcyon)* and Black Phoebe also have a close affinity with watercourses and riverbanks, and though neither is common, both are expected throughout the mountains from the yellow pine belt downslope.

Plate 104. The Yellow-rumped Warbler is found throughout the state but chooses to breed almost exclusively in the upper branches of trees in mountain forests.

Although kingfishers occur on both slopes, phoebes are confined to the western slopes and are quite rare in the more arid mountain regions. Waterfowl are rare along mountain streams, more so now than formerly. Wood Duck *(Aix sponsa)* breeds sparingly in the oak woodland zone of the foothills, and occasionally up into mixed ponderosa pine forest where Pileated Woodpecker holes or human-made nest boxes provide nesting cavities. The Mallard *(Anas platyrhynchos)* has held on tenaciously, and Common Merganser *(Mergus merganser)* still finds breeding possible along large streams at lower elevations.

Southern Sierra at Kern River Valley

At the southern terminus of the Sierra Nevada, the Kern Plateau straddles an intersection between several bioregions—the Sierra Nevada, the San Joaquin Valley, the Great Basin, and the Mojave Desert. Elements of each of these environments exist here, none to the exclusion of the other. As a result, many species that are otherwise absent from the Sierra show up here. The South Fork of the Kern River, perched at 2,600 feet above the San Joaquin Valley and surrounded by southern Sierra peaks, includes 5,000 acres protected by Audubon–California's Kern River Preserve, Sequoia National Forest's South Fork Wildlife Area, and California Department of Fish and Game's Canebrake Ecological Reserve. Astounding for a noncoastal site, 327 bird species have been recorded to date, 197 of which have been documented as nesting locally. Many waterbirds otherwise rare or absent from the Sierra occur regularly in the Kern River Valley; these include American White Pelican *(Pelecanus erythrorhynchos)*, Black-crowned Night-Heron *(Nycticorax nycticorax)*, Blue-winged Teal *(Anas discors)*, Lesser Yellowlegs *(Tringa flavipes)*, Whimbrel *(Numenius phaeopus)*, Marbled Godwit *(Limosa fedoa)*, Bonaparte's Gull *(Larus philadelphia)*, Caspian Tern *(Sterna caspia)*, and Black Tern *(Chlidonias niger)*. Likewise, landbirds generally thought of as lowland species—Yellow-billed Cuckoo *(Coccyzus americanus)*, Lesser Nighthawk *(Chordeiles acutipennis)*, Costa's Hummingbird *(Calypte costae)*—reach higher elevations here, although the nighthawk and hummingbird also range up into

Plate 105. Unlike many other cuckoos, the Yellow-billed Cuckoo is not a brood parasite, but builds its own nest and tends its young. Its breeding success has been severely impacted by the wholesale destruction of willow–cottonwood forests.

Plate 106. A nestling Yellow-billed Cuckoo displays marks in its bill lining, or gape, that are probably directional indicators to assist the parent birds in feeding the chick. Yellow-billed Cuckoo young grow rapidly, developing feathers and leaving the nest within five to six days after hatching.

some desert mountains. Great Basin and desert species also approach their elevational and distributional limits here, mostly because of its proximity to their normal habitats: Ladder-backed Woodpecker *(Picoides scalaris)*, Gray Flycatcher, Brown-crested Flycatcher *(Picoides scalaris)*, Pinyon Jay, Cactus Wren *(Campylorhynchus brunneicapillus)*, Summer Tanager *(Piranga rubra)*, Black-throated Sparrow *(Amphispiza bilineata)*, and Scott's Oriole *(Icterus parisorum)*. More northerly species such as Blue Grouse, Pileated Woodpecker, and Winter Wren approach the southern limit of their year-round distribution. The overlapping habitats that characterize the Kern River Valley are best expressed in an article in the magazine *Birding:*

> Visitors to the Kern River Valley and the southern Sierra Nevada will be fascinated by the interface of so many species pairs. Here, one may find Indigo and Lazuli Buntings breeding side-by-side, Cassin's and Plumbeous Vireos in the same upland forests, and Nuttall's and Ladder-backed Woodpeckers in the same lowlands. (Barnes and Steele 2003)

Raptor Migration

In the same area, Kelso Creek, a tributary of the upper Kern River, runs through a narrow valley that separates Scodie Mountain, a part of the southern Sierra, from Piute Mountain. The area is a funneling point, a bottleneck for bird of prey migration, particularly Turkey Vultures. Apparently, the migrating vultures follow the western slope of the Sierra southward and concentrate here as they cross over a low point in the Sierra and head southeasterly to the Mojave River, which they follow through the desert.

Along the Kern River near the town of Weldon (as at Tehachapi Pass, nearby) an average of 3,000 vultures are seen daily from late September to early October (the same timing of peak raptor migration as on the coast at Marin Headlands—"Birds of the Coast Ranges") with average estimates ranging from 25,000 to 33,000 vultures passing from September through October. Of course, among the throngs of vultures many other raptors are also passing by; up to 16 raptor species have been seen at this spot. The timing of this impressive passage also coincides with peak landbird migration, an added enticement for the avid birder and curious naturalist alike. The annual "Vulture Count" began in 1994 and is scheduled for the last week of September every year, a must-do field trip.

BIRDS OF THE GREAT BASIN

To the casual traveler, driving from Bridgeport southward to Bishop, or across the Modoc Plateau from Alturas westward to Dorris, the expanses of sagebrush desert and alkali flats may seem monotonous, almost infernal, especially in summer. But the snowcapped peaks of the eastern Cascades and the Sierra Nevada, the immutable slopes of Shasta, or the austere western flanks of the White Mountains lend a serene if distant beauty to the landscape. For the intrepid naturalist who ventures out of the car across the sagebrush *playa,* or to the shore of any of the basin lakes, the Great Basin is an exhilarating landscape to explore. It rivals any California region for the wildlife it supports, in particular for its sheer numbers of swans, geese, and waterfowl (especially in early spring), wintering raptors, and breeding sage-grouse and shrubland sparrows.

Map 10. Great Basin Bioregion.

Physical Environment

The Great Basin was named by John Fremont in 1843 upon his realization that there was no outlet to the Pacific. Also called the Intermountain West, the region extends northward into Canada and southward into Mexico, but projects into California only slightly at its western edge. Within California, the Great Basin bioregion lies entirely east of the Sierra–Cascade Range axis, reaching northward to the Oregon border; to the south it merges with the Mojave Desert near the southern Inyo Mountains. In the American west, the Great Basin has a huge expanse, 250,000 square miles, covering much of Nevada, Utah, Wyoming, and Idaho, but within California the Basin covers the least land area of any of the major bioregions. Its extent within the state conforms to the distribution of the Black-billed Magpie *(Pica hudsonia)*, and it is divided into two regions: the Modoc Plateau, which includes the Klamath Basin, Surprise Valley, and the Warner Mountains, and the Eastern Sierra, which includes the Mono Basin and the Owens Valley, the White-Inyo Mountains, and Sweetwater and Glass Mountains, each a unique ecosystem in its own right.

The Great Basin is an arid region, isolated from marine influences, and more allied with the Intermountain West than with other California bioregions. Pacific air masses, moving in from the California Current, are relieved of their moisture first by the Coast Ranges and foothills, and then by the Sierras and the Cascade Range. The Basin lies in the rain shadow of these mountain systems and, as in the Mojave and Sonoran Deserts, very little moisture is available. Northward of the 37th parallel, California's Great Basin occurs mostly above 4,000 feet in elevation, a desert environment with extremely cold winters and hot, dry summers. Average precipitation is low, eight to 15 inches annually, arriving mostly as winter snow. In fact, if it were not for the snowmelt received from the adjacent Sierras or Cascades, the Basin would experience an annual water deficit because precipitation is exceeded twofold by evapotranspiration. Temperatures vary widely, falling well below freezing for much of the winter (average minimum temperatures eight to 15 degrees F) and the summer is "hot to very hot" (Hickman 1993) with average temperatures reaching well into the 80s and 90s.

Biological Environment

The habitats of the basin includes cold desert sagebrush steppe, montane and riparian woodland, basin lakes and associated wetlands, and dry alkali flats, each of which supports its unique community of birds. Sagebrush steppe is the characteristic and dominant plant community, however pinyon–juniper woodland and riparian habitats are relatively common above the alluvial playas, and "islands" of conifers occur at higher elevations. Extensive yellow pine forests occupy the saddle between the Mono Basin and the Owens Valley, and alpine conifer communities are found in the Sweetwater Mountains and on Glass Mountain, and the famous bristlecone pine *(Pinus longaeva)* forests in the White and Inyo Mountains.

The sagebrush steppe may seem to the casual visitor like a monoculture of sagebrush *(Artemesia tridentata)*, but to the careful observer it is a remarkably complex environment. Other bushes are scattered amid the sage—the yellow-flowered rabbitbrush (*Chrysothamnus* spp.), bitterbrush *(Purshia tridentata)*, and desert peach *(Prunus andersonii)* are among the most common, and the sparsely vegetated ground can be covered by a biologically active layer of microorganisms and a profuse array of wildflowers and bunchgrasses.

Above the sagebrush playas, on steeper slopes and rockier soils, pinyon–juniper forests develop, usually relatively open also, with sagebrush and associated shrubs and open ground beneath. Single-leaf pinyon *(Pinus monophylla)*, "the most common tree in the Great Basin" (Stuart and Sawyer 2001), tends to dominate to the southward (Mono and Inyo Counties), giving way to juniper (*Juniperus* spp.) farther northward and at higher elevations.

Zonation of conifer forests at higher elevations, that is, in more *mesic* microenvironments, is similar to the forests of the Sierras and Cascades, with open yellow pine *(Pinus jeffreyi* or, less commonly, *P. ponderosa)* forest often abutting sagebrush flats, and denser mixed-conifer forests of yellow pine, white fir *(Abies concolor)*, and lodgepole pine *(Pinus contorta* subsp. *murrayana)* on higher or damper slopes. Lodgepole pines may dominate on nutrient-poor soils or at old burn sites and high meadow edges.

Riparian woodlands—verdant and dense forests of quaking aspen *(Populus tremuloides)*, black cottonwood *(P. balsamifera)*, and willows (*Salix* spp.)—border perennial watercourses and provide important habitat for breeding as well as migrating birds. Meadows and grasslands tend to be limited in extent, but occur within the sagebrush community where soils and water availability allow, or as *seral* communities following fire. Sparsely vegetated alkali flats have their own complement of tolerant shrubs (e.g., *Sacrobatus* spp.), and relatively barren ground occurs at high elevations, above the xeric, shrubby habitat dominated by mountain-mahogany (*Cercocarpus ledifolius*) and associated bushes, grasses, and *forbs.*

Ecology

Like so many other regions of the American west, the ecological integrity of the Great Basin has been compromised by invasive plant species, which have prospered under a regime of government-subsidized grazing and forestry practices that favor extractive resource management over conservation. Invasive species tend to disrupt the interactions of indigenous species, and in so doing convert diverse communities with complex architectures and refined relationships into structurally simple systems with less complex relationships. The result is fewer species. This dumbing down of ecosystems is termed "ecological relaxation" by ecologists and is equivalent to recession in economic systems. If unabated, recession devolves into depression and economic collapse; likewise, if unattended, ecological relaxation progresses into extinction and ecological collapse. The Great Basin, like so many other dynamic ecosystems in the west, is currently relaxing ecologically.

The rate of conversion from sagebrush to nonnative grassland is not easy to estimate; however, based on the original extent of steppe vegetation (240,000 square miles) and its current extent (less than half the original), it must have averaged several square miles a day over the past century. The native sagebrush desert is comprised of bushes of various ages, sizes, and varieties depending on interrelated environmental variables such as soil structure, moisture availability, elevation, and the history of fire and grazing regimes. Because of the lack of moisture, shrubs tend to be widely spaced; in unaltered sagebrush steppe, bunchgrasses and wild forbs grow in the spaces between the dominant shrubs. A fair share of bare ground also tends to occur in a sagebrush desert, or seemingly barren ground that is actually covered by a diverse mixture of lichens, algae, mosses, and other inconspicuous plants. This complex and often overlooked component of the sagebrush community is called the "cryptogamic crust."

As it happens, this fragile layer of life at the soil's surface plays a critical role in sagebrush steppe stability. The crust aids in the retention of available moisture, contributes nutrients, promotes the establishment of sagebrush seedlings, and prevents invasion by alien plants. The crust also harbors insects and seeds that provide food sources for ground-foraging birds, reptiles, and mammals. So, it functions as a protective shield and also as habitat with its own inherent value. When the cryptogamic crust is disrupted—by trampling, by irregular fire events, by off-road vehicle tracks, or by other unnatural disturbances—the opportunity is ripe for alien species to gain a foothold; as a result, the habitat value is diminished.

Cheatgrass *(Bromus tectorum)* is perhaps the most pernicious weed that is invading sagebrush steppe. Cheatgrass's aggressive nature and adaptation to frequent fires allow it to outcompete native grasses and increase its foothold. Because of the summer aridity of the Basin and the frequency of "dry" lightning, wildfires have become fairly common. In a less-disturbed environment, fires did not burn over large areas because the sparse plant cover and the broken terrain limited their spread. (A study of 110 years of fire history in the Mono Basin found most fires burned fewer than 10 acres and none more than 100 acres—National Science Foundation 1987). The common sagebrush species of California's Great Basin *(Artemisia tridentata)* does not sprout after fire, and the sagebrush ecosystem as a whole is not particularly fire-adapted. Cheatgrass is, however, and because it is an annual and dies back during summer, the period when dry lightning is most likely, it provides a ready fuel; in fact, it could be said to promote fire. As the extent of cheatgrass increases, so does the likelihood of wildfire.

Although conversion of intact sagebrush communities may reduce or eliminate sagebrush obligates, such as Greater Sage-Grouse *(Centrocercus urophasianus)* and Sage Thrasher *(Oreoscoptes montanus)*, some sagebrush steppe birds do occur in shrub-grasslands that have been invaded by cheatgrass and other annual nonnative grasses—Long-billed Curlew *(Numenius americanus)*, Western Meadowlark *(Sturnella neglecta)*, and Lark Sparrow *(Chondestes grammacus)*, for example. However, the invasion and replacement of the sagebrush steppe by cheatgrass may be just a transitional step in an ongoing process. Cheatgrass may be a seral species responding to an initial disturbance; in the future it may be outcompeted by even more pernicious invasives, such as star thistle *(Centaurea solstitialis)* or medusahead grass *(Taeniatherium caput-medusae)*, if management practices are not changed to tilt the balance in favor of sagebrush.

Plate 107. The elaborate display of the male Greater Sage-Grouse is performed to attract females to the lek, where mating takes place.

Sagebrush Steppe and the Greater Sage-Grouse

To understand the complexity of an ecosystem such as the sagebrush steppe, it is useful to consider the life history of a species that is fully dependent on that ecosystem, and how that species may use the habitat at different seasons or during different phases of its life cycle. The Greater Sage-Grouse may be considered an "umbrella species" of the sagebrush community, because its needs encompass those of other species that are dependent on the same habitat type. (It should be noted, however, that the umbrella species concept has been questioned as to its effectiveness at protecting all species, especially invertebrates [Rubinoff 2001].) Sage-grouse biology gives insight into the variable conditions that exist, and are necessary, to sustain itself and other sagebrush-dependent birds. "Their requirements for lek sites, nesting, brood-rearing, and wintering habitat are reasonably well understood. Further, they need large blocks of sagebrush, as much as 2,500 square miles per population, in appropriate spatial mixes across the landscape" (Rich and Altman 2000).

In mid-March, when the sagebrush is still dusted in snow and the steppe glistens with frost, male sage-grouse gather on traditional strutting grounds, called "leks," to begin the breeding season. Lek sites are established in relatively open areas surrounded by sagebrush, display arenas where the big males can strut their stuff, and attract females in from the surrounding plains. The site cannot be too open; surrounding bushes must be available for escape and protection from predators. A lek site might be located in a burn area, a small lake bed or playa, an opening on a ridge, or even on a road. The lek is the center of activity for a local sage-grouse population, and the same site is used year after year as long as the population is healthy and viable.

The ritualized breeding display of the male grouse is one of the most remarkable performances of a North American bird. In first light, or even earlier, the cock commences booming, producing a loud, resonant, percussive sound that can carry long distances across the landscape, attracting females from miles around. The sound is produced by the expulsion of air from two large esophageal sacks on his chest in concert with body pumping, wing beating, and tail fanning. Another rasping sound is made by the male rubbing his fore wing against bristly feathers on his air sacs. The leks may contain dozens of birds, each defending a small territory within the arena; the concert they produce is at once unique, impressive, and bizarre.

Like all lekking bird species, sage-grouse are promiscuous and polygynous. Hens are attracted into the strutting grounds and assume the copulatory position in front of the male who has managed to impress her most with the elaborateness, or outrageousness, of his display. There is some benefit to holding the innermost territory within the lek, a position usually held by the older, more experienced cocks. After copulations, the hen retreats to her nest site, which is generally within a few miles, but may be more than 10 miles from the lek. The female incubates the eggs and broods the chicks independently of the males. For the first few weeks of life, the chicks forage among the sagebrush for insects. As the season progresses, the brood begins to move farther afield—perhaps to a riparian forest, marsh, or field edge—in search of herbaceous plants (forbs, not grasses) on which to feed. In fall and through winter, a sage-grouse depends almost entirely on the leaves of sage for forage.

Greater Sage-Grouse populations can be migratory or nonmigratory (resident). California populations are apparently resident; however, they move seasonally within their broad home range after breeding, up to higher elevations as the snow melts, temperatures increase, and plant growth advances. Nonmigratory populations may spend the entire year within an area of 40 square miles or less in size, depending on the quality and diversity of the local habitat. In more migratory populations, there may be more than one home range area: a breeding range, a brood-rearing range, and a winter range.

Sagebrush steppe habitat has been either lost or degraded through much of the west over the past more than 100 years by a variety of causes: agricultural conversion; broad herbicide treatment; plowing; reseeding of disturbed lands with exotic grains (wheatgrass and cheatgrass); and livestock management practices, including fencing, spring development, and ensuing changes in vegetation and the destruction of the biological crust that contributes to sagebrush health. Fire has become a common event in sagebrush steppe habitat, but the recovery of the habitat is dependent on a sequence of influences following fire. If grazing pressure is too intense too soon after the fire, the living layer of soil, and ultimately the regeneration of the sagebrush, is reduced and the value of the habitat to sagebrush-dependent species is diminished.

Sage-grouse populations have been studied more closely than those of other sagebrush-dependent birds; however, we can infer that all have experienced similar disruptions with changes in their habitat. Throughout the American west, sage-grouse populations have declined by about one-third over the last 30 to 40 years. As a result, the species has been extirpated in five states and one Canadian province, and is at risk in six other states (including California) and two Canadian provinces. No single cause has

Plate 108. Loggerhead Shrikes feed their newly hatched young on spiders and insects, progressing to small rodents and reptiles as the nestlings grow.

been identified for the decline; rather, it is the result of the cumulative effects of the land-use practices described above, which have disrupted the ecosystem.

Other birds most intimately related to the sagebrush community include Sage Thrasher, Brewer's Sparrow *(Spizella breweri)*, and Sage Sparrow *(Amphispiza belli)*. Species that have wider habitat affinities but whose populations include significant proportions of sagebrush steppe habitat and would benefit from protection of sagebrush steppe include Burrowing Owl *(Athene cunicularia)*, Nuttall's Common Poorwill *(Phaleaenoptilus nuttallii nuttallii)*, Horned Lark *(Eremophila alpestris)*, Loggerhead Shrike *(Lanius ludovicianus)*, Spotted Towhee *(Pipilo maculatus)* and Green-tailed Towhee *(P. chlorurus)*, Black-chinned Sparrow *(Spizella atrogularis)*, Lark Sparrow, and Vesper Sparrow *(Pooecetes gramineus)*, among others.

The umbrella species concept is not perfect, of course. Tracts of habitat that support grouse but not thrashers, or vice versa, may exist; however, it seems judicious to assume that if all the complexity of sagebrush steppe is hospitable to the grouse, it also serves the needs of many other species.

The lion's share of the habitat that grouse occupy in California and the rest of the American west is managed by the Bureau of Land Management. Until recently, management of the sagebrush communities for native species has not been a high priority of the Bureau or of any other public agency. As these agencies, the scientific community, and the public have come to understand the dynamics of this environment, all have begun to work cooperatively to develop management plans that recognize the importance of umbrella species, such as sage-grouse, in the long-term health and integrity of the functioning ecosystem. Partners in Flight, a consortium of agency and scientific organizations, has developed cooperative management plans to address the needs of such species and to consider the implications of a wide range of management tools and practices on the various species that rely on a given habitat.

Above the sagebrush plains, the shrub community mixes with conifers, and the density and complexity of the vegetation varies with microclimate, *slope aspect,* and soil characteristics. These transition zones tend to be biologically rich areas (see section "Edge Effect and Ecotone") where species from both higher and lower elevations find the edge of their comfort zone. Here, too, as mountain slopes break and drainages converge, riparian communities become broader and provide more cover for nesting and roosting sites for a variety of species. The richness of the Great Basin–Sierra ecotone is expressed in the abundance of rodents that occurs here, and the corresponding abundance of raptors that prey on them. In an earlier part of this book, we discussed the boom-and-bust cycles experienced by lowland raptors—White-tailed Kite *(Elanus leucurus)* in particular—that specialized on meadow voles (*Microtus* spp.). Microtine rodents also occur along the Great Basin–Sierra ecotone, but so do a variety of other rodents, many of which are long lived and support much more stable populations than voles, and therefore a more reliable prey base. Deer mice (*Peromyscus* spp.), pocket mice (*Perognathus* spp.), kangaroo rats (*Dipodomys* spp.), as well as ground squirrels and chipmunks, are all common.

Long-eared Owl *(Asio otus)* is particularly common in this ecotone. In fact, Grinnell and Miller (1944) considered the Great Basin among the centers of abundance of this wide-ranging but rarely encountered owl. A recent study of owls of Glass Mountain, which lies between the Sierra and the White-Inyo Mountains in Mono County, found relatively high numbers of Long-eareds and attributed that abundance to the rich and reliable prey base provided by rodents (Shuford and Fitton 1998). In the same study, the distribution of small owls found breeding on Glass Mountain provides insight into the elevational and prey relationships one might expect to find throughout the Basin. Northern Saw-whet Owl *(Aegolius acadicus)*, for example, occurs from about 6,000 to 9,800 feet and occupies "a wide range of forests and woodlands of varying age, stature, and openness." Flammulated Owl *(Otus flammeolus)* was associated with yellow pine and aspen groves, mostly at about 8,000 feet. Flams are insectivorous owls, not rodent eaters, and their habitat preferences indicate some insect availability in summer at those elevations. Western Screech-Owl *(Megascops kennicottii)*, a carnivore and insectivore, occurs here at somewhat lower elevations, not associated with yellow pine–riparian habitat, but rather with pinyon pine forest. Northern Pygmy-Owl *(Glaucidium gnoma)* tends to occur lower still, and is very rare in the Basin, but has been found in aspen–conifer edge on Glass Mountain. (Northern Pygmy-Owl is an enigmatic species; it seems to be relatively uncommon through much of its range, except perhaps in the drier mixed evergreen forests of the Coast Ranges and lower Sierra.)

Owls are not the only species of interest in the Great Basin mountains, of course. Other bird families also divvy up the habitat according to their own habits and needs. For example, the common corvids include Black-billed Magpie, Common Raven *(Corvus corax)*, Pinyon Jay *(Gymnorhinus cyanocephalus)*, Western Scrub-Jay *(Aphelocoma californica)*, and Steller's Jay *(Cyanocitta stelleri)*.

Although migratory species such as the owls discussed above are capable of wide-ranging dispersal (otherwise those species would not occur in isolated mountain ranges within the Great Basin), the cordillera of the Sierra Nevada is an effective distributional barrier to species that are not as adept at long-distance flight, especially over high elevations. Several species pairs (congeners) illustrate the evolutionary influence that geographic isolation has played in speciation. As we discussed in "Birds of the Central Valley and Delta," the Yellow-billed Magpie *(Pica nuttalli)* is restricted to the Central Valley and Coast Ranges south of San Francisco Bay. In the Great Basin, the Black-billed Magpie is a common bird of the open sagebrush community, and ventures upslope to about 8,000 feet. These two magpie populations probably became isolated and diverged in the Pleistocene, and their distinctive bill color and behaviors now distinguish two different species. The Oak Titmouse *(Baeolophus inornatus)* and the Juniper Titmouse *(Baeolophus griseus)*, likewise, have been separated on opposite slopes of the mountain ranges, and in their isolation have developed different habitat associations, as their common names suggest. The titmice are relatively bottom-heavy, or sedentary species, a characteristic that affords them little opportunity to interbreed, each busy in its own habitat niche in tracts of forest that meet only in a very limited contact zone, where oak and juniper intermix, of the Cascade Range in the Klamath Basin.

Another example of a sibling species complex is the Pacific-slope Flycatcher *(Empidonax difficilis)* and the Cordilleran Flycatcher *(E. occidentalis)*. Like the titmice, these two species are also separated by the ridge of the Sierras and the Cascades; the latter species is an endemic member of the Great Basin avifauna. But unlike the titmice, these *Empidonax* flycatchers are highly migratory, spending their winters in the same southern latitudes, and probably coming into contact on their wintering grounds. Because their breeding grounds (and probably the timing of their migrations) are separate, the two populations have developed and maintain a degree of genetic independence that qualifies them as separate species.

Another *Empidonax*, the Gray Flycatcher *(E. wrightii)*, is a Great Basin species with a strong affinity for sagebrush. The Gray Flycatcher deserves mention here to give you the reader some insight into the importance of behavior and habitat niche over morphology in the speciation process, at least for some groups. This complex of small flycatchers, the Empidonaces, which includes seven western species, are so similar in appearance to human eyes that even the most experienced field ornithologists become befuddled by their identification, especially when encountered out of habitat during migration. Like the others, the Gray is highly migratory, wintering in Central America. It returns to the Basin in May, and sets up territory amid sagebrush, or where pinyon forest is thin and open.

Common Nighthawk *(Chordeiles minor)* is one of the latest spring arrivals to the Basin (late May to early June), a function of its affinity for high-flying insects (moths) that are on the wing on warm summer nights. Nighthawks nest directly on the ground, and are cryptically colored to blend with the cryptobiotic crust that covers open areas in undisturbed sagebrush steppe. Nighthawks occur not only on the basin plain, however, but at high elevations, as in the White-Inyo Mountains. Their breeding display, a booming aerial ballet, competes with those of Wilson's Snipe *(Gallinago delicata)* and Short-eared Owl *(Asio flammeus)* as one of the most dramatic aerial performances to be witnessed on a summer night in the basin. (The nighthawk booming should not be confused with that of the sage-grouse, a decidedly earthbound dance that serves an entirely different mating system; the nighthawks are monogamous and monomorphic, whereas sage-grouse are promiscuous and dimorphic.)

Basin Wetlands

The mountain ranges within the Great Basin are high enough to create their own weather patterns and, though dry, do gather moisture from the parched air. The snowy peaks of the eastern Sierra, Mount Shasta, and the Cascades, and the ranges wholly within the basin provide meltwater that runs down from the higher slopes into drainages that feed basin lakes. The most notable basin lakes, and those that support the greatest abundance of birdlife, include Mono Lake, Eagle Lake, Honey Lake, Goose Lake, and the several wetlands within the lower Klamath Basin. Each has its own unique attributes and its own seasonal rhythms and cycles, mostly dependent on the amount of runoff available and the water levels attained:

> The interior lakes constitute a series of disjunct and intermittently-available oases, whose users must be semi-nomadic and capable of shifting as environmental conditions require. . . . The health of bird populations that use unstable habitats is to a large extent dependent on back-up sites that can be used when conditions change. Unfortunately, there is no redundancy left in the saline and alkaline lakes of the west. (Jehl 1994)

Here, because of their individual complexity and diversity, we discuss two of the most important basins—Mono Lake and the lower Klamath—and mention the salient features of several others, but leave further discovery to the curious reader.

Mono Lake

One of the most historic and heroic environmental battles of the twentieth century took place between the Los Angeles Department of Water and Power and the tiny but tenacious Mono Lake Committee. The story was a classic David and Goliath battle, and fortunately for the birds, David won. David Gains, in fact, was one of the first birder-biologists-visionaries to recognize the value of Mono Lake to California's birds; he helped found the Mono Lake Committee, which took the cause to the public and, ultimately, to the courts. With tremendous public support backed by sound science, the Committee was able to document the damage that Department's diversion of water would exact on the Mono Lake ecosystem and the public trust. The courts agreed, and the greed that had been draining the life from the Mono Basin was abated.

Now, Mono Lake is designated a Western Hemispheric Shorebird Reserve of international importance (one of three in the Great Basin, along with the Great Salt Lake in Utah and the Lahontan Valley in Nevada) because 30 species of shorebirds use the lake during spring and fall migration. Numbers are highest in fall, when American Avocets *(Recurvirostra americana)*, Red-necked Phalarope *(Phalaropus lobatus)*, and Wilson's Phalarope *(Phalaropus tricolor)* are particularly abundant. The lake also supports California's largest breeding colony of California Gulls *(Larus californicus)*, is an important inland breeding site for the Western Snowy Plover *(Charadrius alexandrinus)*, and is a critical migratory staging area for Eared Grebes *(Podiceps nigricollis)*.

> At saline and alkaline lakes, birds are often conspicuous top consumers in relatively short food chains. Birds are attracted to lakes such as Mono because they provide an unusual abundance of food. The harsh chemical conditions of the lake preclude the existence there of many kinds of aquatic predators, particularly fish. Without aquatic predators, populations of some aquatic invertebrates reach extraordinarily high densities. (National Science Foundation 1987)

At Mono Lake, the invertebrates that occupy the base of the food chain are brine shrimp *(Artemia monica)* and brine fly *(Ephydra hians)*. The life histories of the most abundant birds that exploit these resources elucidate interactions within a relatively simple system.

California Gull

At Mono Lake, California Gulls nest in vast colonies on islands within the lake. This nesting area is inaccessible to predators, especially coyotes, that would otherwise decimate the colony. One of the primary concerns regarding the Department's drawdown of the lake was the creation of a land bridge that would allow predators to prey on the gull colony, an event that actually occurred in 1979, when water levels became critically low. Because of a State Supreme Court decision in 1983 supporting the Public Trust doctrine, and a decision by the State Water Resources Control Board in 1994 to limit the Los Angeles Department of Water and Power's drawdown and maintain the lake level at 6,392 feet, the lake has persevered as viable wildlife habitat.

The critical period for any species is when it is raising its young. California Gull chicks begin hatching in late May and continue through June. Although gulls are notorious omnivores, the Mono Lake gulls rely largely on brine shrimp during the

Plate 109. Important foods for California Gulls that nest on islands in Mono Lake are brine shrimp and alkali flies. The gulls have found an effective way of feeding on flies by scooping them up as they run through swarms with open beaks.

chick-rearing period of their breeding cycle. The brine shrimp population at the lake reaches its peak during June, providing a reliable and timely resource for the breeding gulls. In the mid-1990s, careful surveys revealed about 24,000 to 31,500 breeding pairs of gulls at Mono Lake, representing an astounding 70 to 80 percent of the state's entire population (Shuford and Ryan 2000).

Competition for the shrimp is at a minimum during the early summer; other birds that exploit this food source, such as phalaropes and grebes, do not arrive at the lake until later in the season.

Phalaropes

Toward the end of the gull's breeding season, phalaropes arrive at Mono en route from their breeding grounds to wintering grounds farther south. Although all three species of phalaropes have occurred at Mono, Red Phalarope *(Phalaropus fulicarius)* is essentially an oceangoing migrant and occurs only as a vagrant. Wilson's Phalarope is the most abundant of the three, beginning to arrive in late June after breeding; numbers peak in August and diminish through September. Peak numbers upward of 70,000 are sustained by swarms of both brine shrimp and brine flies. Mono Lake is apparently the largest staging area for Wilson's Phalaropes in the western states. The birds stay for an extended period, molting feathers and building fat reserves in preparation for the long southbound migration to wintering grounds in Argentina. Energized by Mono reserves, some individuals make the long journey in as few as three days (Jehl 1994)! Large numbers of Red-necked Phalaropes begin arriving in mid-July, after breeding far to the north, and remain into early September. Red-neckeds occur only in about half the numbers of Wilson's, and their stay is shorter. Red-necked Phalarope is apparently a pickier eater, selecting brine flies over brine shrimp. It winters mostly in the tropical Pacific, but some also winter at scattered sites in San Diego and Imperial Counties.

Eared Grebes

The dent in the prey base made by the gulls and phalaropes is not enough to deter Eared Grebes from congregating at Mono Lake in fall, when well over a million birds occur at once, representing one-third or more of the North American population. Like Wilson's Phalarope, the grebes stay for an extended period, gaining

Plate 110. Golden ear tufts mark the summer plumage of an Eared Grebe.

body weight and replacing body feathers. Eared Grebes apparently select brine shrimp as their primary prey, especially the shrimp larvae that are abundant in fall. Most grebes leave the Basin in late fall, migrating by night in large flocks. In his book, *Birds of the Great Basin,* Frank Ryser (1985) recounts an amazing incident in which Eared Grebe migration was interrupted by inclement weather in December 1928; grebes fell out of the sky, littering the ground over a 65-mile front centered around Caliente, Nevada. Eared Grebes are also gregarious in their breeding behavior, nesting in colonies in the Great Basin; Eagle Lake, near Lassen Volcanic National Park, supports California's largest aggregation.

As mentioned above, Mono Lake also provides nesting habitat for the Western Snowy Plover; about 10 percent of the California population nests on the exposed lakeshore. Plovers also nest on the beaches of the outer coast (see "Birds of the Shoreline"); however, those populations are suffering progressive declines, thus adding to the importance of the more stable inland breeding sites. Snowys do not forage on the brine shrimp, but do take flies as well as invertebrates associated with freshwater springs and seeps, especially along the lake's eastern shore. California Gulls exact a toll on plover chicks, apparently, but overall this population has been reported as viable (Page et al. 1983).

Plate 111. A Western Snowy Plover broods its newly hatched, precocial young on the shores of Mono Lake. The sandy lake shoreline has proved to be an important breeding ground for this endangered plover.

The Mono Lake Basin, lying at the boundary of the Sierra Nevada and the Great Basin, with elevations ranging from about 6,380 to 13,000 feet, holds overlapping species characteristic of both bioregions. More than 300 species of birds have been reported from the Mono Lake Basin, representing all the major groups of California birds, and about 118 species nest here. Highly pelagic species are poorly represented, of course, but such a large body of water attracts even oceanic wanderers, far off course, but undoubtedly relieved to find a semblance of the sea. As mentioned above, Red Phalarope has occurred on several occasions; more surprising, however, are records of all three species of scoter, three jaegers, and Sabine's Gull *(Xema sabini)*. Mono Lake also holds one of the few California records of Long-billed Murrelet *(Brachyramphus perdix)*, a vagrant from Asia! Surprisingly, pelagic birds have been found at most of the larger Great Basin water bodies, a curious phenomenon.

About half of the birds that occur around the Mono Basin are relatively uncommon, a fact that increases the excitement of birding the area. The list of the Mono Basin includes not only vagrant waterbirds, but also vagrant landbirds such as Eastern Kingbird *(Tyrannus tyrannus)*, Bohemian Waxwing *(Bombycilla*

Plate 112. The flutelike song of the Western Meadowlark asserts its territorial claim.

garrulus), Chestnut-sided Warbler *(Dendroica pensylvanica)*, and Bobolink *(Dolichonyx oryzivorus)* alongside expected ones such as Western Kingbird *(Tyrannus verticalis)*, Cedar Waxwing *(Bombycilla cedrorum)*, Yellow Warbler *(Dendroica petechia)*, and Western Meadowlark.

The Owens Valley to the south of the Mono Basin formerly held Owens Lake, a hypersaline wetland 110 square miles in area. Owens Lake is to the Great Basin as Tulare Lake is to the Central Valley; an environment destroyed before much was known about its ecological value. Newspaper accounts from the late 1800s do reveal that "whole navies of aquatic birds" arrived at the lake in late summer, and it is safe to assume the avifauna was similar to that currently using Mono Lake. The Owens River, which fed the lake, was diverted to supply Los Angeles in 1917, and the lake died (Jehl 1994). Recent attempts to irrigate the dry playa, however, have led to a revitalization of the basin floor; birds—particularly migrant waterfowl and shorebirds—have begun returning to the area. The Owens Valley is an important inland breeding site for Snowy Plover, and Long-billed Curlew has bred here in recent years. Cottonwood riparian forests have survived in the Owens Valley, and conservation efforts are under way to increase their

Plate 113. The Whip-poor-will makes no nest, laying its eggs on leaf litter in open forest. During incubation it relies on its cryptic plumage to cover and conceal the light-colored eggs, and it "sits tight" rather than flushes when danger threatens.

extent, to the benefit of many riparian-obligates such as Swainson's Hawk *(Buteo swainsoni)*, Yellow-billed Cuckoo *(Coccyzus americanus)*, and Willow Flycatcher *(Empidonax traillii)*. Although degradation has characterized the history of the Owens Valley, it appears that restoration may characterize the future.

The White-Inyo Mountains rise steeply from the eastern edge of the Owens Valley, extending into the southeastern corner of California's Great Basin, a unique ecosystem unto itself. Its complexity is beyond the range of this book, so the interested reader is referred to *Natural History of the White-Inyo Range, Eastern California* (Hall 1991) for a thorough treatment of these mountains. Suffice it to say that several species breed there, particularly on the eastern slopes, that are rare elsewhere in California's Great Basin—Whip-poor-will *(Caprimulgus vociferus)*, Broad-tailed Hummingbird *(Selasphorus platycercus)*, and Virginia's Warbler *(Vermivora virginiae)* among them. Scott's Oriole *(Icterus parisorum)*, primarily a bird of the Mojave Desert, breeds in the White-Inyos, an indication of the area as a transitional zone between two bioregions.

Klamath Basin

Geologically, the Klamath is within the volcanic Southern Cascade–Modoc Plateau bioregion; however, because of similarity of climate and habitat, we include the area in the Great Basin. Tucked up against the Oregon border, and bound to the west by the Cascades, the Klamath Basin is perhaps the best place in North America to see concentrations of waterfowl. Tundra Swan *(Cygnus columbianus)*, Snow Goose *(Chen caerulescens)*, Ross's Goose *(C. rossii)*, and Greater White-fronted Goose *(Anser albifrons)*, and dabbling ducks are particularly well represented. These legions of waterfowl attract, in turn, numbers of Bald Eagles *(Haliaeetus leucocephalus)* to the area in winter, several hundred of which visit the ponds or fields during the day and retreat to a communal roost at night. One of the "must-dos" for birders visiting the Klamath Basin is to witness the Bald Eagle "fly-out" on a cold February morning. At first light, hundreds of eagles leave canyon roosting sites to fly out into the flatlands in search of injured or dying waterfowl. Concentrations of the eagles tend to be highest on cold nights in February, when counts of 400 to 600 eagles have been tallied on the annual Bald Eagle Festival centered in the town of Dorris.

The Klamath Basin extends northward into Oregon (beyond the scope of this book), but the most productive waterfowl areas—lower Klamath, Tule Lake, and Clear Lake National Wildlife Refuges—are within California. Midwinter numbers of swans and geese moving among the "water units" are impressive and may give an inkling of what California was like before the arrival of Europeans. Unlike the Mono Basin, an essentially natural water body, the hydrology of the Klamath is highly modified. The system of levees that creates shallow-water habitat is managed for waterfowl and agricultural water use, and has been so controlled for nearly 100 years. Historically, the Klamath Basin encompassed 185,000 acres of shallow-water and marsh habitat and supported an estimated six million waterfowl during autumn peaks. More than 75 percent of the wetland habitat was lost when the area was converted to agricultural lands.

The Klamath Basin is linked to the Central Valley refuges (see "Birds of the Central Valley and Delta"). Waterfowl flocks, most noticeably Snow Geese, arrive in the Basin in mid-October by the tens of thousands. They stay until open water freezes over, and

Plate 114. Bald Eagles are attracted to the vast aggregations of wintering waterfowl in the Klamath Basin. Birds that succumb to the cold are an appealing food source for the eagles.

then proceed down to the Sacramento Valley refuges to spend midwinter. In late winter or early spring, as the days lengthen, temperatures rise, and ice melts, the flocks shift back northward into the Klamath. The smaller Ross's Goose is relatively uncommon in the Klamath in fall, but returns with the flocks of Snows in early spring, swelling the throngs of white geese. By March, the daily movement of Snows and Ross's from the refuge roosting ponds into the agricultural foraging fields is a startling, sublime sight. Mixed together with white-fronted geese, these flocks can surpass 100,000 birds. Tundra Swans also overwinter in the Basin with large numbers (about 15,000) arriving in mid-October and remaining through March.

On a clear day in late winter or early spring, the sight of throngs of immaculate white geese and swans winging across the basin, with the snowy backdrop of Mount Shasta and the southern Cascades, is as exhilarating and beautiful a natural spectacle as you can hope to see anywhere on earth.

Winter is the season of greatest avian abundance, but the managed wetlands support waterbirds year-round. Common breeding species include Pied-billed Grebe *(Podilymbus podiceps)*, Eared Grebe, Western and Clark's Grebes *(Aechmophorus*

FEBRUARY 22, 2003. KLAMATH AND TULELAKE NWRS.

About 5:30 a.m., without a hint of morning light, we walk up a frosted gravel road near Bear Valley National Wildlife Refuge to see the dawn Bald Eagle flyout. A Northern Pygmy-Owl calls steadily from the juniper; a Mountain Quail whistles. We hear the whooshing of their wings before we can see the birds. The sky lightens incrementally, and visibility gradually improves. We begin to see their muscular forms, following the ridge, flying from the communal night roost up the forested canyon out toward the basin, singly or in groups of no more than three birds. Soon the light is bright enough to notice the emblematic white head leading the massive dark body and those broad, oarlike wings. By 7:30 we've counted more than a hundred bald eagles, and the procession begins to dwindle to a few stragglers.

We follow the eagles out to the basin to where they've dispersed, roosting on the levees among the water units, watching the flocks of waterfowl.

At White Lake, well over a thousand Tundra Swans are gathered, foraging and resting on the still water. The reflection off the surface of the cerulean sky and the swans makes the flock seem double its actual size. A few days earlier, someone had reported a rare Whooper Swan among these throngs. We search the flocks for a larger swan with a larger yellow base to its bill. In the distance, toward the shining silhouette of Shasta, a circling eagle flushes a flock of Snow Geese from another unit. The swans become alert, with necks stretched, the chatter, a kind of throaty garbling, intensifies. Then the geese are fully aloft, several thousand flying westward. The sense of calm has been breached, some collective consensus reached, and now the swans, too, lift off the water, their powerful legs kicking and wings beating, roiling the water's surface, shattering the mirror effect of the previously still water. Other flocks arise from other lakes, and soon the sky is stippled with birds—waterfowl of all sorts—shifting among the units, searching for forage and tranquility.

By day's end we estimated that we'd seen 20,000 Snow Geese and Ross's Geese, 5,000 Tundra Swans, intermixed with a few thousand wigeon, shoveler, pintail, and scaup. Tomorrow we'll return to find the Whooper Swan, but this day, invigorated by the sheer abundance of life, will remain with us, in memory and spirit, for life.

Plate 115. Western Grebes breed on inland waters such as Eagle Lake. Their dramatic courtship ritual begins with the pair bowing to each other before rising in tandem to sprint across the surface.

occidentalis and *A. clarkii)*, Double-crested Cormorant *(Phalacrocorax auritus)*, Great Blue Heron *(Ardea herodias)* and Great Egret *(A. alba)*, White-faced Ibis *(Plegadis chihi)*, Canada Goose *(Branta canadensis)*, Mallard *(Anas platyrhynchos)*, Cinnamon Teal *(Anas cyanoptera)*, Redhead *(Aythya americana)*, and Ruddy Duck *(Oxyura jamaicensis)*. The Klamath contains the state's last breeding colony of American White Pelican *(Pelecanus erythrorhynchos)* (Clear Lake and Sheepy Lake), a species that was historically much more widespread in California and the west. This is also an important breeding ground for Ring-billed Gull *(Larus delawarensis)*, California Gull, and Forster's Tern *(Sterna forsteri)*. Common Nighthawks are common breeders (present only in summer), as are, of course, many species of passerines. Sheepy Ridge, a long escarpment that parallels Tule Lake's west shore shelters numerous nesting raptors—Red-tailed Hawk *(Buteo jamaicensis)*, Prairie Falcon *(Falco mexicanus)*, American Kestrel *(Falco sparverius)*, as well as Common Raven.

The Great Basin in general, and the Klamath in particular, attracts a few northerly species rare elsewhere in the state; Northern Shrike *(Lanius excubitor)* and American Tree Sparrow *(Spizella*

Plate 116. The Clark's Grebe has a bright yellow bill with the white of the face pattern extending above the eye, which serves to distinguish it from the Western Grebe. The two were long considered one species and do occasionally interbreed.

arborea), for example, are fairly regular visitants. Harris's Sparrow *(Zonotrichia querula)* and Snow Bunting *(Plectrophenax nivalis)* occur irregularly; and Gyrfalcon *(Falco rusticolus)* is a rare possibility. Conversely, some passerine species common in lowland California are relatively rare here, such as American Crow *(Corvus brachyrhynchos),* Black Phoebe *(Sayornis nigricans),* Western Bluebird *(Sialia mexicana),* and Northern Mockingbird *(Mimus polyglottos).*

For all its wild nature, the Klamath region has been the focus of intense environmental battles in recent years. Unlike that of Mono Lake, the future of the Klamath Basin is not secure. The system of irrigation ditches and dams that transformed the original sagebrush steppe and wetlands of the Klamath Basin into productive farmland was created by the U.S. Bureau of Reclamation in 1902 to attract settlers to these arid lands. The Klamath Lake National Wildlife Refuge was established in 1908, the first refuge in the national wildlife refuge system (created by President Theodore Roosevelt in 1903) dedicated to waterfowl. Clear Lake was established in 1911, and Tule Lake National Wildlife Refuge in 1928. In average- or high-precipitation years, there is enough water to go around. When there is a

Plate 117. American White Pelicans nest in colonies on islands in inland lakes (often in the company of Double-crested Cormorants), laying their eggs in simple depressions. Colonies are sensitive and may be abandoned if disturbed.

drought or a limited snowpack, two government-subsidized beneficiaries—agriculture and wildlife—are inherently at odds. Agricultural interests are ever thirsty for the water allocated to sustain the refuges and the Klamath River salmon run. In dry years, the competition is fierce; half of the years between 1990 and 2002 approached drought conditions, and the tension between wildlife needs and agricultural desires intensified. No resolution seems forthcoming.

Bird populations have suffered in the Klamath as elsewhere. Numbers of geese and wildfowl that spend the winter in the refuges are holding their own, but in lower than historical numbers and with less habitat available in dry years. Historically, six million waterfowl crowded into the basin; today about one-third of that number occur during peak periods. Many of the basin's breeding species are less stable. Greater Sage-Grouse numbers have dwindled with loss of sagebrush steppe habitat, and only one lek persists, at Clear Lake National Wildlife Refuge. Predictably, Burrowing Owl declines have paralleled those of the grouse, and it may have been extirpated as a breeding species. Snowy Plover, too, has declined, with only one breeding location still active at lower Klamath. According to the U.S. Fish and

Wildlife Service, "113 out of 410 wildlife species identified in the Klamath Basin are considered to be of concern or at risk" (www.onrc.org/programs/klamath/wildlife.html).

Like Mono Lake, the Klamath region attracts its share of vagrants, even some surprising oceanic birds: Pacific Loon *(Gavia pacifica)*, Brown Pelican *(Pelecanus occidentalis)*, Black Turnstone *(Arenaria melanocephala)*, Red Phalarope, and Black-legged Kittiwake *(Rissa tridactyla)*. Perhaps more anticipated, some exotic waterfowl have come in with the throngs: Whooper Swan *(Cygnus cygnus)* and possibly "Bewick's" Swan *(Cygnus columbianus bewickii)* have been picked out of the flocks of Tundra Swans; Emperor Goose *(Chen canagica)*, Baikal Teal *(Anas formosa)* and Garganey *(A. querquedula)*, Harlequin Duck *(Histrionicus histrionicus)*, and Black Scoter *(Melanitta nigra)* have all occurred.

Modoc Plateau

The volcanic region of the Modoc Plateau, in the northeasternmost corner of the state, is an area unique unto itself, but it shares affinities with the Great Basin and the southern Cascades. Habitats include sagebrush steppe and grassland, basins and alkali flats, lava flows, mountains, and important water bodies, both freshwater—Eagle Lake and Goose Lake—and the hypersaline Honey Lake. The Warner Mountains and Surprise Valley are also considered to lie within this bioregion. Much of the most productive bird habitat is included within the Modoc National Wildlife Refuge, the Modoc National Forest, and Lassen Volcanic National Park.

The northwestern plateau was formerly the haunt of the Columbian Sharp-tailed Grouse *(Tympanuchus phasianellus columbianus)*, but sadly, no longer. This relatively unpeopled region, with expanses of open space, still supports a rich and varied avifauna. It is perhaps the best region for raptors in both winter and summer. Golden Eagles *(Aquila chrysaetos)* are as common as anywhere in the state, and Eagle Lake supports substantial breeding colonies of both Bald Eagles and Osprey *(Pandion haliaetus)*. Swainson's Hawk has been extirpated as a breeding species from the southern coast and has declined by an estimated 90 percent in the Central Valley, so the northeast, where declines

Plate 118. Ring-billed Gulls flock over their breeding ground at Honey Lake.

have not been as severe, has become an important refuge for this raptor, and about 20 percent of the state's population of breeding pairs occurs here. Butte Valley, near the Klamath National Wildlife Refuge, is a particular stronghold. (The southern Owens Valley is also important.) Northern Goshawk *(Accipiter gentilis)* is resident in the Warner Mountains, and the adjacent Surprise Valley may be the most likely place in California for Gyrfalcons to appear, in fall or winter, although none of the state's several records have been from here. Flammulated Owls, too, are fairly common in the Warners (e.g., Jess Valley).

The wetlands—reservoirs, impounded creeks, marshy grasslands, and saline lakes—of the Modoc Plateau support the greatest diversity of breeding waterfowl in California, not only ducks but also Eared, Western, and Clark's Grebe, White-faced Ibis, Sandhill Crane *(Grus canadensis)*, Ring-billed Gull, and Black Tern *(Chlidonias niger)*, Forster's Tern, and Caspian Tern *(Sterna caspia)*. Eagle Lake hosts the state's largest breeding colonies of Eared and Western Grebes.

Goose Lake is an important staging area for Canada Goose and Canvasback *(Aythya valisineria)* in late summer and early fall. Vast flocks of shorebirds, particularly American Avocets, and Wilson's Phalaropes stage at various shallow-water sites in the

Plate 119. A Forster's Tern offers food to its mate in a courtship ritual common among tern species.

region in late summer and fall, depending on water levels: Goose Lake and Honey Lake host exceptional concentrations. Honey Lake rivals Mono Lake as a critical migration stopover for shorebirds and as a breeding area for Snowy Plovers. Honey Lake formerly supported a significant American White Pelican breeding colony, but that was abandoned when the lake began drying because of agricultural drawdown in summer months; sadly, this pattern is being repeated at various sites throughout the west. The surrounding sagebrush sea still supports an active sage-grouse lek, but, as elsewhere, this may be imperiled.

The Big and Fall River valleys, on the western edge of the Modoc Plateau, near the junction of Shasta, Lassen, and Modoc Counties, are the transition zones between the southern Cascades and the Plateau. This is one of the few places in California—along with Eagle Lake and Lake Tahoe—where Bufflehead *(Bucephala albeola)* breeds. Thousands of "Greater" Sandhill Cranes *(Grus canadensis tabida)* visit in spring en route from their Central Valley wintering grounds to northerly breeding grounds, and a few dozen stay to breed. Snow, Ross's, and Cackling Canada Geese *(Branta canadensis minima)* also stage here in large numbers; the latter is a diminutive and rather rare subspecies of the abundant honkers that are on the

Plate 120. Sandhill Cranes require large expanses of isolated marshland in which to breed. Incubating cranes relieve each other periodically in a ceremony called "change over," which is often accompanied by loud and dramatic trumpeting.

Plate 121. A female Northern Harrier prepares to drop onto the nest with food for her young. The food has been caught by the male and, in an impressive aerial maneuver, passed to the female in flight.

increase throughout the state. Nearby McArthur-Burney Falls Memorial State Park is a well-known birders' stop, particularly for the Black Swift *(Cypseloides niger)* colony that breeds behind the falls.

Plate 122. A Northern Harrier feeds morsels of food to young in her nest hidden in tall weeds.

Plate 123. Small groups of Black Tern breed in loosely dispersed colonies, each pair building an island of aquatic vegetation that keeps the nest dry and predator-free.

Butte Valley Wildlife Area, Siskiyou County, is an important breeding ground for five larids: California and Ring-billed Gulls and Black, Forster's, and Caspian Terns. Swainson's Hawks are holding on in good numbers here, and as in much of the area, Bald Eagles are common in winter.

The South Warner Wilderness has interesting pairs of species, such as Williamson's Sapsucker *(Sphyrapicus thyroideus)* and Black-backed Woodpecker *(Picoides arcticus)*, Red-naped Sapsucker *(S. nuchalis)* and Red-breasted Sapsucker *(S. ruber)*, Calliope Hummingbird *(Stellula calliope)* and Rufous Hummingbird *(Selasphorus rufus)*, and *Empidonax* flycatchers. Gray-crowned Rosy-Finches *(Leucosticte tephrocotis)* have occurred in autumn.

Plate 124. A newly fledged Short-eared Owl presents its fearsome-looking facial display to deter potential aggressors.

Surprise Valley is a raptor hangout in fall and winter, and the town of Fort Bidwell is the most reliable place in the state to find Northern Shrike and Bohemian Waxwing, two California rarities. The breeding ranges of a few species rare elsewhere in the state extend into this northeastern corner, notably Bobolink and Eastern Kingbird.

THE DESERTS' BIRDS

The wise traveler will always get up early in the morning with the rock wrens and catch the changing glories of the coming day. Also, he will spend the closing hours of the day in quietness so as . . . to hear, in contrast to the intense stillness of evening, the clear, liquid notes of that magnificent bird singer of the mesquite country, the LeConte Thrasher . . . remarkable for its richness of tone.

EDMUND JAEGER, *THE CALIFORNIA DESERT*

The Physical Environment

The North American Deserts are divided into four types based on their vegetation: the Great Basin, the Chihuahuan, the Mojavan, and the Sonoran. The Chihuahuan lies fully outside California, in Arizona and New Mexico. The northernmost, coldest, high-elevation Great Basin was covered in the previous chapter. The Mojavan and Sonoran types are middle- and low-elevation deserts, respectively. Together they encompass 154,000 square

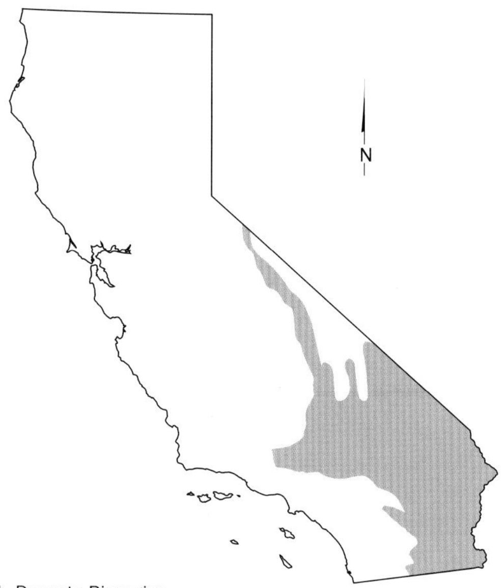

Map 11. Deserts Bioregion.

miles of California arid lands, reaching from the Owens Valley in the north to the Mexican border in the south and east of the Sierra and the Peninsular ranges to the Nevada and Arizona borders. Of California's three arid provinces, two intrude into the state from much larger areas. The Great Basin extends in from the sagebrush steppe and Great Plains to the north and east. The Sonoran extends in from the southeast and Mexico; that portion of the Sonoran within California is known as the Colorado Desert. The Mojave Desert is essentially an area of transition between the Colorado Desert and the Great Basin, with a mix of birds from each region (Miller 1951).

Roughly one-third (21 million acres) of the state's deserts are covered by creosote bush *(Larrea tridentata)*, "the most abundant shrub in California" (Schoenherr 1992), an indication of this plant's ability to tolerate the drought stress imposed by one of the earth's most challenging environments. Because its seeds require several years of cool and moist weather to germinate, they may lie dormant in the desert ground, perhaps a holdover from their origins under a Pleistocene climate. Alternatively, creosote

Map 12. Geographic subregions of the deserts.

can propagate from an underground crown and produce clones, literally genetic replicas of itself. Judged from the apparent uniform age of shrubs on the creosote plain, and relative rarity of young individuals, cloning is the more common reproductive strategy. Indeed, individual creosote plants have been aged at 11,000 years or more; these individuals have been alive since the Pleistocene, and so are among the oldest living plants on earth.

The Mojave and the Sonoran Deserts differ greatly climatically and floristically, and different species of reptiles and mammals tend to be associated with one or the other, but many of the same birds occur in both. In a faunal study of Joshua Tree National Park, located at the southern extreme of the Mojave and encompassing the transition zone into the Sonoran Desert, Miller and Stebbins (1964) found that "Birds as a whole tend to be less narrowly limited than reptiles in a zonal sense, or in terms of altitudinal plant belts, in this area." That said, some bird species are more closely affiliated with one region over the other, and those affinities are considered in this chapter.

The Mojave Desert

The Mojave is the smallest of California's deserts, lies entirely within the state, and extends from the southern Owens Valley southward to the Little San Bernardino Mountains. The transition zone between the Great Basin and the Mojave Desert is relatively wide (50 to 60 miles) and complex, a mix of smaller transition zones and habitat edges, with changes in elevation and climate driving changes in the biota. Some general characteristics are clear, however. As you move southward from the Great Basin into the Mojave, elevation decreases and temperature increases gradually. Whereas the Great Basin receives most of its moisture as snow, the Mojave receives most of its as rain. As sagebrush is the characteristic shrub of the Great Basin, creosote bush is the characteristic shrub of the Mojave.

The Mojave lies at middle elevations of about 2,000 to 4,000 feet above sea level. At higher elevations, as in Joshua Tree National Park, snow is a rare but annual event. Annual temperatures in the Mojave are highly variable, with summer highs averaging 100 to 110 degrees F and winter lows averaging 30 to 42 degrees F.

Daily temperatures can also swing wildly, presenting a daunting challenge for desert wildlife. Rainfall decreases from west (about 10 inches per year) to east (two inches per year) and occurs mostly in winter. The western Mojave, therefore, is influenced somewhat by coastal climate. This overlap of coastal and desert influences supports a diverse mixture of habitat types and, consequently, bird communities. The western edge of the Mojave, along the base of the Peninsular Ranges, supports pinyon forest that extends down into the Mojave and attracts birds that are largely absent from the Sonoran Desert. The avian community of this pinyon–yucca ecotone includes Western Scrub-Jay *(Aphelocoma californica)*, Pinyon Jay *(Gymnorhinus cyanocephalus)*, Oak Titmouse *(Baeolophus inornatus)*, Bushtit *(Psaltriparus minimus)*, Blue-gray Gnatcatcher *(Polioptila caerulea)*, Bewick's Wren *(Thryomanes bewickii)*, Gray Vireo *(Vireo vicinior)*, Spotted Towhee *(Pipilo maculatus)*, and Black-chinned Sparrow *(Spizella atrogularis)*, most of which (except Pinyon Jay) are also birds of the coastal slope and chaparral. Species that range upward from the low-elevation desert to the lower edge of the pinyon–yucca–juniper forest of the mountain slopes include Gambel's Quail *(Callipepla gambelii)*, Mourning Dove *(Zenaida macroura)*, Western Screech-Owl *(Megascops kennicottii)*,

Plate 125. Black-throated Sparrows are the most widely distributed of desert sparrows, occurring throughout the southern deserts and Great Basin.

Greater Roadrunner *(Geococcyx californianus),* Ladder-backed Woodpecker *(Picoides scalaris),* Costa's Hummingbird *(Calypte costae),* Ash-throated Flycatcher *(Myiarchus cinerascens),* Crissal Thrasher *(Toxostoma crissale),* Scott's Oriole *(Icterus parisorum),* and Black-throated Sparrow *(Amphispiza bilineata)* (Cardiff and Remsen 1981).

"The desert claims the finest-singing and the handsomest of California's orioles. The trim black-and-yellow Scott oriole is a regular resident of the yucca forests of the Mohave Desert, and occasionally it is found flashing its brilliant form among the junipers" (Jaeger 1933).The smell of creosote after a spring rain is as characteristic and as evocative of the Mojave as a Scott's Oriole singing from a yucca flower, or a desert tortoise plodding across the road. Creosote-covered *bajadas* can appear monotonous, but other shrubs are interspersed, commonly burroweed *(Ambrosia dumosa)* and burrobrush *(Hymenoclea salsola).* The fruits of creosote and burrobrush have evolved prickers to adhere to fur and feathers, and so are dispersed by animals. Below about 3,500 feet in elevation, the creosote community occurs on well-drained soils, especially on the broad alluvial fans that comprise a vast portion of the Mojave. On less-well-drained saline and hardpan soils, a shadescale scrub community occurs, characterized by saltbush (*Atriplex* spp.), hop-sage *(Grayia spinosa),* winter fat *(Krascheninnikovia lanata),* and Mormon tea (*Ephedra* spp.). Shadescale scrub tends to be shorter and denser than creosote, and many shrubs are thorny. As in the Central Valley and the Great Basin, an alkali scrub community occurs in the most poorly drained, most alkali or saline soils; most plants are halophytes in the goosefoot family (Chenopodiaceae).

Like the plant community, the avian community of the Mojave is not exceptionally diverse, but its members are a handsome lot. Black-throated Sparrow is perhaps the most obvious Mojave resident, and, though widely distributed throughout desert regions, is most closely allied with the Mojave. Its numbers swell in fall by birds retreating southward from the Great Basin; by midwinter most birds leave the open desert, or flock along its western edge. Black-throateds have adapted to the intense heat of the Mojave by seasonally varying their diet. In winter, when water is available, they are primarily seedeaters; in summer, when water is unavailable, they switch to insects and green vegetation, which provide enough water for survival. Black-throateds may join the

Plate 126. The Gambel's Quail is well adapted to its hot, dry habitat. It augments a winter diet of seeds with insects and vegetation to gain moisture in the drier months.

mixed flocks of White-crowned Sparrow *(Zonotrichia leucophrys)*, Brewer's Sparrow *(Spizella breweri)*, and Sage Sparrow *(Amphispiza belli)* that arrive in winter, but in the spring and summer breeding season they are strictly territorial.

The breeding distribution of Scott's Oriole, one of the North America's most striking birds, is also allied with the Mojave, especially the Joshua tree and yucca woodlands, but also upslope into pinyon–juniper forests of the desert mountains, where other members of the yucca genus—Our Lord's candle *(Yucca whippelei)* and Mojave yucca—grow. Scott's tends to "sew" its nest to the edges of dead yucca leaves. Scott's are strictly summer residents, arriving in mid-April and departing for the tropics in mid-July. The North American orioles—Scott's Oriole, Orchard Oriole *(Icterus spurius)*, Bullock's Oriole *(Icterus bullockii)*, and Baltimore Oriole *(Icterus galbula)*—are migratory and highly dimorphic, with the males much more brightly plumaged than the females. Interestingly, in many of the nonmigratory tropical orioles, males and females are very similar in appearance.

Gambel's Quail, a desert version of the more familiar California Quail *(Callipepla californica)*, "symbolizes the Sonoran Desert

Plate 127. A male Costa's Hummingbird, the most common hummer of the arid deserts, has a distinctively elongated, amethyst-colored gorget.

almost as much as the saguaro cactus and gila monster" (Brown et al. 1998). It also occurs in the lower reaches of the Mojave (e.g., Death Valley), but is largely absent north of Joshua Tree. California and Gambel's Quail, a species pair that constitute a superspecies, come into contact at San Gorgonio Pass, at the western edge of the Mojave (and in xeric chaparral communities of San Diego County), where they hybridize rarely. They also come into contact in Baja California (northward of San Felipe), where hybridization has been suspected but not verified. When areas of contact between two species are very limited and stable, and when interbreeding is confined to a very small percentage of individuals, taxonomists usually recognize the two species as distinct (Omland 2001). Gambel's Quail and Mountain Quail *(Oreortyx pictus)* also occur together, but at higher elevations (upper elevations of the Little San Bernardino Mountains), especially in the vicinity of springs in the nonbreeding season, but have not been found to hybridize.

Doves, like quail, require water to complement their granivorous diets and are not particularly well adapted to desert life. Doves overcome this limitation by flying long distances across the desert floor to water sources. Both Mourning Dove and White-winged Dove *(Zenaida asiatica)* occur in the Mojave, but Mourning Dove is far more numerous, abundant even. White-wingeds reach the northern extent of their distribution in the southern Mojave, increasing southward into the Colorado Desert.

In riparian areas of the Mojave, cottonwoods occur, especially in oases, but mesquite, with its deep reaching roots, becomes the characteristic species. Catclaw *(Acacia greggii)* and velvet ash *(Fraxinus velutina)* are common associates, and in saline situations iodine bush *(Allenrolfea occidentalis)* is common. In highly degraded riparian systems, the alien tamarisk *(Tamarix chinensis)* or "saltcedar" has taken over, and the animal communities are altered in favor of ecological generalists over the more diverse and ecologically refined native community that previously existed. Tamarisk forests tend to be colonized by blackbirds and House Finches *(Carpodacus mexicanus),* replacing native desert riparian species such as Bell's Vireo *(Vireo bellii),* Lucy's Warbler *(Vermivora luciae),* and Summer Tanager *(Piranga rubra).*

The Joshua tree *(Yucca brevifolia)* is the only "tree" of the Mojave, and it is widespread above the basin floor, with the most impressive stands at Cima Dome and Joshua Tree National Park. Ladder-backed Woodpecker is a common resident of the Joshua tree forest and an emblematic bird of the California deserts. Lad-

Plate 128. White-winged Doves are summer visitors to the desert, coexisting with Mourning Doves.

Plate 129. The female Ladder-backed Woodpecker lacks the red crown of the male. This species is very similar in coloration to the Nuttall's Woodpecker, and its call, a series of rapidly descending notes, is like that of the Downy Woodpecker.

derbacks come in contact with the similar and closely related Nuttall's Woodpecker *(Picoides nuttallii)* of the oak woodland at a few places where coastal and desert influences collide, such as Morongo Valley and Yaqui Wells in Anza-Borrego Desert State Park. (Hybridization may occur at these points of contact.) The soft wood of the Joshua tree also provides ready access to other cavity-nesting species such as American Kestrel *(Falco sparverius)*, Northern Flicker *(Colaptes auratus)*, and Ash-throated Flycatcher. Western Screech-Owl sometimes nests in Joshua trees, though sparingly; in Anza-Borrego, Screech-Owls are most closely associated with fan palms *(Washingtonia filifera)*.

Common Poorwill *(Phalaenoptilus nuttallii)* requires attention because the circumstances of its discovery tell a story of unique adaptation to the Mojave climate. Poorwills eat insects, a food source of variable availability in the cold Mojave winters, when daily temperatures can swing more than 50 degrees F. Whereas other insectivores, such as flycatchers, swallows, and warblers, migrate south to avoid cold periods, the poorwill is the first bird that was found to use hibernation as a way to deal with cold and the lack of food availability. Edmund Jaeger, a pioneer desert naturalist, discovered a *torpid* poorwill in the Chuckwalla Mountains, east of the Salton Sea, on a cold December in 1946. Torpidity was suspected for this species, but had never been documented. The bird returned to the crevice in the same granite wall the following winter and the ever-curious Jaeger also returned, refound the bird, and took its temperature on several occasions over the course of the winter (Jaeger 1948). Since Jaeger's experiments, torpidity in swifts and hummingbirds has been thoroughly studied. White-throated Swifts *(Aeronautes saxatalis)* roost and nest in cliff cracks and crevices of exfoliating granite and are present throughout the year, but active only when temperatures rise enough to allow insects to take wing. Like poorwills and hummingbirds, swifts have the ability to go torpid when temperatures drop, and so reduce their metabolic demands to withstand periods of starvation.

Wide-ranging species to be expected throughout the desert regions include Mourning Dove, Lesser Nighthawk *(Chordeiles acutipennis)*, Greater Roadrunner, Loggerhead Shrike *(Lanius ludovicianus)*, Say's Phoebe *(Sayornis saya)*, Costa's Hummingbird, Rock Wren *(Salpinctes obsoletus)* and Cactus Wren *(Campylorhynchus brunneicapillus)*, Black-throated Sparrow, and the ubiquitous House Finch. Less common, locally occurring birds with restricted ranges are Bendire's Thrasher *(Toxostoma bendirei)* and Crissal Thrasher, Lucy's Warbler, Summer Tanager, and Great-tailed Grackle *(Quiscalus mexicanus)* (though expanding its range). Of the large desert flycatchers, two—Ash-throated Flycatcher and Western Kingbird *(Tyrannus verticalis)*—are fairly obvious and occur throughout, and two others—Brown-crested Flycatcher *(Myiarchus tyrannulus)* and Cassin's Kingbird *(Tyrannus vociferans)*—have restricted ranges and spotty distributions.

Common Ravens *(Corvus corax)* are a common sight throughout the Mojave, nesting in cliffs and remote outcroppings, but wandering widely across the bajadas and patrolling highways for roadkill. Other aerial members of the Mojave security force include Red-tailed Hawk *(Buteo jamaicensis)* and Golden Eagle *(Aquila chrysaetos)*.

Cima Dome, a low-relief, huge granitic bulge, is located within the federally managed Mojave National Preserve and supports an extensive stand of Joshua trees with a scattering of creosote bush and grasses amid the forest. The grasses are a remnant of former grazing practices; the area was managed by the Bureau

Plate 130. The Common Raven is an opportunist that finds easy pickings at highway rest stops. It is also a resourceful predator that poses a threat to the eggs and young of nesting birds.

of Land Management before being designated as a National Scenic Area in 1994. This area has long been a destination for birders because three rather elusive thrashers—Crissal Thrasher, Bendire's Thrasher, and Leconte's Thrasher *(Toxostoma lecontei)*—all occur in the area, and from mid-March through summer, Cima is the best place in California to find Bendire's Thrasher. Other birds of the high-elevation Mojave such as Rock Wren and Scott's Oriole are also conspicuous members of the Joshua tree community.

Mountains of the Deserts

The mountains of the eastern Mojave—most notably Clark Mountain, and the Kingston, New York, Providence, and Granite ranges—which cluster near the Nevada border in eastern San Bernardino County, are an archipelago of islands in a creosote sea. The Pleistocene plant communities that inhabit these mountain islands are refugees from an earlier epoch and contrast starkly with the surrounding desert. The Pleistocene Epoch, which spanned a period from 1.8 million to about 11,000 years ago, was cooler and wetter during its latter phase than our modern era, and communities supported by that climate have retreated to the higher elevations as the deserts have advanced. Steep elevational changes characteristic of the Mojave, and the gradation of climatic conditions that they produce, support zones of habitat above the surrounding playas (bajadas). Joshua tree forest (as at Cima Dome) is replaced above by pinyon–juniper forest intermixed with hard chaparral. Higher still, conifer forests of white fir *(Abies concolor)* or limber pine *(Pinus flexilis)*, which used to extend farther down in elevation, now occur only above 7,000 feet, and are therefore limited in extent.

The higher peaks attract and support breeding bird species, but very limited numbers of individuals, that are at the western limit of their ranges and breed at only a few sites elsewhere in California—Whip-poor-will *(Caprimulgus vociferus)*, Broad-tailed Hummingbird *(Selasphorus platycercus)*, and Hepatic Tanager *(Piranga flava)*. Pioneers or wanderers from the Arizona mountains—Painted Redstart *(Myioborus pictus)*, Red-faced Warbler *(Cardellina rubrifrons)*, and Grace's Warbler *(Dendroica graciae)*—also have been found here, but apparently occur only sporadically. One wonders if formerly, when the habitat was more extensive and less islandlike, these species bred more commonly in these ranges and retain some vague genetic memory that this area is part of their historic homeland. But birds do wander, and they are such a successful class of animals largely because of their ability to find new habitats, or niches within habitats, as these become available.

Broad-tailed Hummingbird, Gray Flycatcher *(Empidonax wrightii)*, and Virginia's Warbler *(Vermivora virginiae)* are associated with the arid conifer and shrub zone. (These species similarly occur in the White-Inyo Mountains, representing a northward extension of the Mojave [see "Birds of the Great Basin"]). Plumbeous Vireo *(Vireo plumbeus)* and Juniper Titmouse *(Baeolophus griseus)* also breed in the pinyon–juniper woodland, two species that were recognized as distinct in 1998. (Plumbeous was formerly considered a race of the Solitary Vireo *[V. solitarius]* complex; "Plain Titmouse" was split into Juniper Titmouse and Oak Titmouse.)

The Argus Mountains in southern Inyo County contain an insular, relict population of California Towhee *(Pipilo crissalis)* (discussed under "endemism" in "Overview of California's Birds").

Plate 131. Bell's Vireos build their nests in scrub-lined washes and are especially vulnerable to the attention of parasitic cowbirds. This has resulted in a steep decline in desert populations.

Big Morongo Canyon, located at the base of the Little San Bernardino Mountains, just north of Palm Springs is, together with the Kern River Valley (see "Birds of Mountains and Foothills"), a premier example of a transition zone, or a mosaic of overlapping habitats, within California, if not North America. It represents a nexus of Mojave, Sonoran, montane, and coastal habitat elements and influences. Here, a perennial watercourse runs out the mouth of Cottonwood Canyon, through Covington County Park and then down through a steep-walled desert canyon. Riparian woodland adjacent to freshwater marsh and alkali meadow, desert shrub with elements of both the Mojave and Sonoran deserts, and oak woodlands intermix within a very small area of several hundred acres. Breeding specialties include some of those with the most restricted ranges in California: Brown-crested Flycatcher, Summer Tanager, Bell's Vireo, Yellow-breasted Chat *(Icteria virens)*. This is also a zone of contact between several closely related species: Gambel's and California Quail, Ladder-backed and Nuttall's Woodpeckers, Ash-throated and Brown-crested Flycatchers.

Water is a rare commodity in the deserts and associated ri-

Plate 132. Largest of the wood-warblers, the Yellow-breasted Chat is more often heard than seen. Despite its striking coloration, it is adept at concealment in the dense thickets in which it is found.

parian habitat is very limited, so those few riparian patches attract and concentrate large numbers and varieties of migrant birds in spring and fall. Well-known vagrant traps associated with the desert mountains include Westgard Pass, Toll House Springs, and Deep Springs Valley, between the White and Inyo Mountains, and Panamint Springs in the Panamint Mountains, Inyo County. Lower elevation migrant traps include Morongo Valley and Whitewater Canyon, Mesquite Springs, Furnace Creek Ranch, and Scotty's Castle in Death Valley, Oasis in southeasternmost Mono, and Fort Piute in San Bernardino County. Butterbredt Spring, located in Kern County on the western edge of the Mojave at 3,800 feet elevation, has a small cottonwood–willow riparian patch supported by perennial surface water. This location at the base of the Tehachapi Mountains makes it a favored stop for migrant passerines and birdwatchers, especially in spring. Almost all of the eastern vagrants that have been observed in California have occurred at these sites, but these vagrant waves occur very rarely, and many hours of searching are usually expended before the lucky birder scores. Nevertheless, the regular species that pass through these oases make for exciting and satisfying birding.

The Colorado Desert

> *Included in it are all of those areas which drain into the Colorado or the Salton Sea. . . . Since the name "Colorado Desert" seems to be a poorly chosen one, it is well to point out that it was given because most of the area lies along the Colorado River and not because of any connection it has to the Colorado River.*
>
> EDMUND JAEGER, *THE NORTH AMERICAN DESERTS*

California's Colorado Desert, a northern subregion of the larger Sonoran Desert, lies south of the Mojave and extends southward across the Mexican border into northern Baja California. Even hotter and drier than the Mojave, it is lower in elevation with most of its area below 1,000 feet and much below sea level. Elements of the Colorado Desert also extend into low-lying reaches of Death and Panamint Valleys, surrounded by a Mojavian environment. The average annual rainfall is about 3.5 inches, about half of which occurs during the summer months as monsoonal fronts move northward out of the Gulf of California. The other half occurs in winter, cast off by Pacific oceanic fronts. Winter temperatures are mild, rarely dropping below freezing. Summers are blistering; extremes above 120 degrees F are recorded regularly at Death Valley and the Imperial Valley.

The arid Colorado Desert occupies much of southern California, including the Coachella, Borrego, Palo Verde, and Imperial Valleys, encompasses the Salton Trough, and extends eastward to the Colorado River and southward into northeastern Baja California and to the northwestern corner of the Gulf of California.

The transition zone between the Mojave and Sonoran is more abrupt than that between the Great Basin and the Mojave. Driving from Twentynine Palms southward through Joshua Tree National Park toward Mecca, you will soon notice the disappearance of Joshua trees. The waving red torches of the ocotillo *(Fouquieria splendens)* signal your descent into the Colorado Desert. Because of the relative warmth of the climate, the Colorado is botanically more diverse than the Mojave. Creosote still occurs, but is less dominant, being joined by many species of succulents mostly absent from the Mojave. The distribution of ocotillo along the gravels of the upper bajada effectively outlines

Plate 133. The Verdin favors mesquite groves and dry thorn thickets, where it gleans insects and their larvae from leaves and buds.

the extent of the Sonoran Desert. Trees also grow in the washes and dry arroyos—mesquite (*Prosopis* spp.), smoke tree (*Psorothamnus spinosa*), blue palo verde (*Cercidium floridum*), desert ironwood (*Olneya tesota*), and catclaw acacia.

A short walk up a Colorado Desert wash soon introduces you to another indicator plant—cholla cactus (*Opuntia* spp.)—which becomes abundant as soon as you cross into the frost-free zone. Here, too, the birds change rather dramatically. The Verdin (*Auriparus flaviceps*), with its yellow head and insistent personality, is one of the most frequently encountered birds in these desert washes, perhaps because it is constantly feeding to maintain its high metabolism and because of its continual calling ("tzeeeeee"). One of the smallest desert species (weight about 0.2 ounces), Verdins apparently get all the water they need from the insects they feed on. According to Weathers (1983), Verdins "have never been seen drinking." The distinctive round nest, a small ball of twigs, is obvious amid the spiny branches of the palo verde and catclaw, which provide protection from predators. Verdins are prodigious avian architects, building several nests per season; the nests may provide a survivable microclimate in the harsh desert environment. The Verdin is a unique species in the North American avifauna, the only member of the penduline tits (of the Old World family Remizidae) in the New World; all other members are restricted to Eurasia and Africa. (Another taxonomic peculiarity of California, the Wrentit *[Chamaea fasciata]*, is the sole representative of the Old World babbler family [the Timaliidae] naturally occurring in North America [see "Birds of the Shoreline"].)

Cactus Wren occupies these same washes, or adjacent slopes, where cholla cactus provides nesting sites. Like Verdins, Cactus Wrens also build extra secondary nests to provide roosting sites; the nests may serve other functions as well, as decoys from predators, perhaps. Sharing a desert wash with the large and aggressive Cactus Wren may not always benefit the Verdin. Wrens have been known to appropriate Verdin nests and may also prey on their young. The behavior of the Cactus Wren exemplifies the toughness of this environment. Curious as to how the wren negotiated the dangers of the cholla spines, Wesley Weathers set up a blind to watch them entering and leaving a nest within a cactus. He noted that when a spine pierced the wren's callused foot, the bird would simply hold the spine with its beak and yank its foot free!

Greater Roadrunner is perhaps the ultimate denizen of the desert, and his oddity has inspired naturalists accordingly: "The Creator has endowed this desert masterpiece with endearing qualities, such as ought to assure him an enduring welcome. . . .

Plate 134. The formidable cholla cactus seems to have little effect on the Cactus Wren, which builds its ball-shaped nest in the densest of spine tangles.

Plate 135. The Greater Roadrunner's distinctive cooing call is reminiscent of that of other members of the cuckoo family, to which it belongs, but this species is unique in that it is essentially a ground-dweller.

Seen on the desert proper, the Road-runner strikes you instantly as being the fitting thing" (Dawson 1923). Unlike the gregarious quail or dove, roadrunners forego their neighbors, pairing for life and remaining year-round on their territories. They are carnivores, or even omnivores. Lizards provide their primary prey, but roadrunners are not picky eaters, as Jaeger reports: "The paisano's appetite is as queer as his looks. . . . He swallows horned toads, grasshoppers, mice, centipedes, millepedes, cutworms, spiders, bumblebees, and occasionally even snakes, baby wood rats, and newborn rabbits" (Jaeger 1961). And we might add carrion. But lizards are a roadrunner's bread and butter, and this most reptilian of birds is more adept than any other—shrike, kestrel, or owl—at securing lizardly fare. Indeed, the "ground cuckoo" has so taken to heart the old canard "you are what you eat" that it has adapted a reptilian way of life. On cold desert mornings, roadrunners position themselves facing away from the sun and erect their back feathers to expose their skin, which is black, to the available solar energy. They absorb the sun's warmth passively and raise their body temperature, and so mimic the behavior of cold-blooded reptiles, their major prey base (Weathers 1983).

It is important to acknowledge that the ground cuckoo has

I remember sitting on a granite outcrop above a palm oasis one April morning early during a migration wave in southern Anza-Borrego. Gambel's Quail, Scott's and Hooded Orioles, Verdin, and Phainopepla represented the locals as a colorful pageant of migrants paraded through from dawn until midmorning, when the sweltering sun subdued their activity. Nashville, Orange-crowned, and MacGillivary's Warblers, Western Tanager, Lazuli Bunting, Lawrence's Goldfinch, Black-headed Grosbeak, Black-chinned Hummingbird, each made its appearance accompanied by the whistles of the the resident Phainopepla and the scolding Verdin.

lost ground at the periphery of its range. Any observation outside the desert is now rare. Roadrunner used to be widespread in the Sacramento Valley, but where it still exists there, it is very isolated, a doomed population. They are very short winged and cannot fly long distances; therefore, dispersal through fragmented habitat is an impossibility. As Ralph Hoffmann predicted in 1927, conversion of chaparral and shrublands to agricultural and suburban sprawl has ignored the roadrunner's needs: "Like the cow-puncher, the horned-toad and the rattlesnake it will not persist unless there is some land too broken for cultivation, where cactus or chaparral still flourishes" (Hoffmann 1927).

Where water is available, often at canyon heads or fault fractures, groves of native California fan palm are clustered in true palm oases. These dependable water sources attract wildlife, particularly birds.

Black-tailed Gnatcatcher *(Polioptila melanura)* is a true desert bird, a year-round resident of dry washes and desert scrub in the Colorado; its range extends northward into the driest portions of the eastern Mojave. This gnatcatcher is particularly associated with honey and screwbean mesquite *(Prosopis glandulosa* and *P. pubescens);* but along the lower Colorado River it uses tamarisk *(Tamarix chinensis),* which may prove to be a saving grace, given the dominance of that invasive plant in formerly desert scrub habitat. This prediction may be overly optimistic, however; several authors have noted this species' absence from nonnative habitats, and it has been identified as "a good indicator species for undisturbed habitats" (Farquhar and Ritchie 2002). Black-tails used to be considered conspecific with what is now called the California Gnatcatcher *(Polioptila californica)*, but then Jonathan Atwood (1988) discovered that the two "subspecies" were largely *allopatric* and that where their distributions do overlap (northeastern Baja California), they are behaviorally and genetically distinct and thus qualify as separate sister species. According to the biological species concept, "geographic isolation leads to genetic change and potentially to reproductive isolation of sister taxa" (American Ornithologists' Union 1998). Subsequent study has found that the two populations do come into contact and may, in fact, display limited interbreeding; nevertheless, "all arid land gnatcatchers (four species) [are] closely related but genetically distinct at the species level" (Farquhar and Ritchie 2002). It should be noted that the westernmost (Sonoran Desert) race of Black-tails *(P. melanura lucida)* that occurs in California is morphologically distinct from the eastern (Chihuahuan Desert) race, and future examination of those disjunct populations may shed more light on their taxonomy. The gnatcatcher complex provides an object lesson in the mechanism of evolution in California's birds, the process by which taxonomic decisions are made, and the confusion that confronts the observer.

As a diminutive inhabitant of Sonoran Desert scrub, the Black-tailed Gnatcatcher is similar to two other small, foliage-gleaning insectivores—Verdin and Lucy's Warbler. Among other similarities (monogamous, *altricial* young, etc.), each is able to derive enough metabolic water from its insect prey to survive the extreme heat of its environment. This is a crucial desert adaptation for a small bird because small size requires a very high metabolism and, therefore, entails a high rate of water loss.

Birds that are most closely allied with the Sonoran Desert include Lesser Nighthawk, Gambel's Quail, Greater Roadrunner, Gila Woodpecker *(Melanerpes uropygialis)*, Ladder-backed Woodpecker, Verdin, Cactus Wren, Crissal and Le Conte's Thrashers, Black-tailed Gnatcatcher, Bell's Vireo, and Summer Tanager. The occurrence of species outside the Sonoran Desert, for example, Greater Roadrunner and LeConte's Thrasher in the San Joaquin Valley, indicates a former connection between these regions. Species that are essentially restricted to the lower Colorado River and occur (or occurred) only marginally in California include Elf Owl *(Micrathene whitneyi)*, Gila Woodpecker, Northern Cardinal *(Cardinalis cardinalis)*, and Abert's Towhee *(Pipilo aberti)*.

Salton Sea and the Imperial Valley

> *Although it was created in 1905 by accidental flooding by water diverted from the Colorado River, it lies entirely within a portion of historical Lake Cahuilla, a vastly larger ephemeral lake that was last filled in the mid-1600s. In this regard the Salton Sea can be viewed as a natural part of the Sonoran Desert ecosystem. . . . From a biogeographic perspective, the Salton Sea and its adjacent valleys behave as a northern extension of the Gulf of California.*
>
> MICHAEL PATTEN, *CHECKLIST OF THE BIRDS OF THE SALTON SEA*

On April 19, 1908, the perspicacious Joseph Grinnell boarded the vessel *Vinegaroon* to survey the Salton Sea's birdlife. He reported small flocks of Western Grebe *(Aechmophorus occidentalis)* and Eared Grebe *(Podiceps nigricollis),* Double-crested Cormorants *(Phalacrocorax auritus),* Caspian Terns *(Sterna caspia),* Ring-billed Gulls *(Larus delawarensis),* and (surprisingly!) a dozen Common Loons *(Gavia immer).* His first stop on "Echo Island" was "exciting in the extreme," because he found a vast nesting colony of 2,000 American White Pelicans *(Pelecanus erythrorhynchos),* the southernmost colony found to date. Farther along, at another island, they found about 150 Double-crested Cormorant and several Great Blue Heron *(Ardea herodias)* nests. So began the visitation to this curious site by generations of curious naturalists, and their findings have been thoroughly summarized in a recent book, *Birds of the Salton Sea: Status, Biogeography, and Ecology* (Patten et al. 2003), to which the curious reader is referred.

The Salton Sea lies currently at 227 feet below sea level and is the largest inland body of water west of the Rockies: about 45 miles long and 16 miles wide, covering about 444 square miles of the Imperial Valley. It is quite shallow (30 to 40 feet deep on average), and about 25 percent more saline than sea water, with salinity levels increasing in summer when temperature levels rise and evaporation increases. It is one of the harshest environments in North America, and the ability of birds to survive there is a testimony to their adaptability. Much of the surrounding landscape is devoted to agricultural practices (500,000 acres in the Imperial Valley), which ostensibly benefits certain species such as White-

Plate 136. The Burrowing Owl hunts mainly during the early evening and night. It is rarely active by day but may often be seen perched on a post or sunning itself near the entrance to a nest hole.

faced Ibis *(Plegadis chihi)*, Cattle Egret *(Egretta rufescens)*, Mountain Plover *(Charadrius montanus)*, and Burrowing Owls *(Athene cunicularia)*; in some seasons, each of these occurs in greater concentrations here than anywhere else in California. The Imperial Valley supports up to 16,000 ibis in winter, as many as 30,000 pairs of nesting Cattle Egrets, and most of the Mountain Plovers wintering in California. Each of these species is a fairly recent arrival, or has recently increased in the Imperial Valley. Cattle Egrets did not breed prior to 1970; ibis and plovers have increased since the early 1990s. An estimate of about 6,500 pairs of Burrowing Owls in the Imperial Valley represents 70 percent of the estimate for the entire state (Desante and Ruhen 1995), where the species is in precipitous decline. Black Terns *(Chlidonias niger)* also gather in large flocks over flooded agricultural lands in fall. The attraction of the agricultural fields to birds may have hidden dangers and long-term consequences, however, as levels of agricultural contaminants (selenium, boron, organochlorines) have been found to be higher in the Imperial Valley than in other irrigated agricultural regions of California.

Because of its recognition as an avian hot spot, the Salton Sea was designated as a Wildlife Refuge in 1930. For all its challenging extremes, the sea attracts vast numbers of birds of more than 400 species and has been called one of the most important wetlands to birds in North America. It may also be one of the most dangerous: "Historically, birds at the Salton Sea have experienced mortality attributed to a variety of diseases including botulism, Newcastle disease, avian cholera, and salmonellosis, and possibly to toxins produced by algae" (Shuford et al. 1999). Indeed, the Salton Sink may be becoming a population sink for pelicans. The American White Pelican colonies found by Grinnell in 1908 grew over ensuing years, but, sadly, this magnificent creature has been extirpated as a breeding species over the last 30 years, mirroring a disturbing pattern throughout California and the west. But impressive flocks of nonbreeders still concentrate at the sea, with high counts of 26,500 to 33,000 individuals in the 1990s (Shuford et al. 1999); most pelican flocks seen soaring over the Colorado Desert are likely headed for its largest water body. Outbreaks of avian botulism have taken the lives of thousands of white pelicans each summer since 1996, when an estimated 9,000 succumbed to the disease. Increasing summer salinity creates ideal conditions for botulism bacteria to thrive, but die-offs have occurred in winter as well. Brown Pelicans *(Pelecanus occidentalis)* have established a small breeding colony only recently (1996), and now they, too, are suffering die-offs. Both pelican species are critically threatened throughout their ranges, and it is a tragic irony that they are attracted to a habitat and then poisoned there. Botulism has also infected 35 species of waterbirds that use the lake in summer: "only birds that consume Tilapia *(Oreochromis mossambicus)*, a saltwater sport fish introduced from Africa, were directly infected by the disease" (Johnson 2000). Double-crested Cormorants and Eared Grebes were most heavily affected. The mortality of 150,000 Eared Grebes in 1992 was from unknown causes, but it foreshadowed the subsequent episodes.

The sea is a place of wildly varying conditions and rapid change. Birds, so sensitive to environmental variability, are good indicators of change. Most species show tremendous variation in abundance between years, so the following numbers are provided only to give a sense of the magnitude of bird use. Of 12 ardeids (egrets, herons, bitterns, and night-herons) recorded at the sea, eight or nine have nested, Cattle Egret most abundantly; the sea also supports important colonies of Black-crowned Night-Heron *(Nycticorax nycticorax)* (upward of 1,300 pairs). Of

MOUNTAIN PLOVER

The problems of trying to delineate species by bioregion is exemplified by the changing distribution patterns of birds in response to changes in habitat and availability of resources. In the early 1900s, large numbers of Mountain Plovers were reported in California on both grasslands and agricultural lands. Because nearly all the native grasslands and plains have been lost in the San Joaquin Valley, this species' range in the Central Valley is likely less than 10 percent of its former extent, despite its adaptation to using disked and irrigated agricultural fields. The Mountain Plover is a bare-ground specialist whose winter distribution is confined almost entirely to California. Historically, Mountain Plovers "were once abundant on the coastal plains and interior valleys from Ventura County to San Diego County, including Riverside County" (Hunting 2003). They are now rare because of habitat conversion from dry, native rangeland to tract housing. Over the last 50 years, they have occurred most reliably in the Central Valley and more recently in the Imperial Valley. In the mid-twentieth century Mountain Plovers were found abundantly in native rangelands and alkali flats in the western portions of the Central Valley (e.g., Carrizo Plain), which they reached by migrating southward along the western edge of the Great Plains, then swinging westward through New Mexico and Arizona and northward into California's Central Valley. In the mid-1940s, the Imperial Valley was developed for agricultural crops, creating suitable habitat that intercepted the plovers en route to their traditional wintering grounds. Now they occur on fallow fields or heavily grazed sites largely devoid of vegetation. This must be a secondary habitat, since their original habitat, now largely obliterated, was California's native valley grassland. The Plover is currently a Bird Species of Special Concern and considered "threatened," because the population has declined so drastically since the mid-1960s. Another short-grass and bare-ground specialist, the Lark Sparrow *(Chondestes grammacus)*, has also increased in the Imperial Valley over the last few decades. Formerly a winter visitor, it now breeds regularly.

five grebes, Pied-billed Grebe *(Podilymbus podiceps)*, Western Grebe, and Clark's Grebe *(Aechmophorus clarkii)* breed regularly. Clark's Grebe is twice or three times as common as Western Grebe, both of which occur year-round. Eared Grebes are abundant in winter.

Plate 137. The Caspian Tern is the largest and most distinctive of terns. It feeds in a typical ternlike manner, patrolling open waters with its large red bill pointed downward, searching for fish.

Double-crested Cormorant is a common permanent resident, breeding in association with Caspian Terns in colonies that contain some of the largest nesting congregations (2,000 to 3,000 pairs of each species) in the state.

For waterfowl, the sea is most important as wintering habitat, second only to the Central Valley (Shuford et al. 1999). Average numbers for each of the five most common wintering species—Northern Pintail *(Anas acuta)*, American Wigeon *(A. americana)*, Ruddy Duck *(Oxyura jamaicensis)*, Green-winged Teal *(A. crecca)*, and Northern Shoveler *(A. clypeata)*—exceed 10,000 individuals, and Lesser Scaup *(Aythya affinis)* and Bufflehead *(Bucephala albeola)* are fairly common. Ross's Goose *(Chen rossii)* and Snow Goose *(C. caerulescens)* also winter in huge flocks (recent estimates of 30,000) and Ross's Goose, unrecorded in the region prior to the 1940s, has increased as a wintering species by "3–4 orders of magnitude over the past 50 years" (Patten et al. 2001, 2003). Breeding ducks are few, as expected given the stress imposed on nestlings by such a saline environment: Mallard *(A. platyrhynchos)*, Cinnamon Teal *(A. cyanoptera)*, Redhead *(Aythya americana)*, and Ruddy Duck are the regular nesters. Fulvous Whistling-Duck *(Dendrocygna bicolor)* breeds here as well, the only known location left in California. Only a few pairs still occur, although it was formerly fairly common; nesting pairs have also disappeared along the lower Colorado River.

The sea supports an interesting nesting assemblage of larids (gulls and terns): the largest nesting colony of Black Skimmers *(Rynchops niger)* (about 500 pairs) in North America, comprising 40 percent of the California population; several hundred Gull-billed Tern *(Sterna nilotica)* represent the only landlocked colony in North America and one of only two colonies on the west coast north of Mexico. (The other is in coastal San Diego County.) California Gull *(Larus californicus)*, Forster's Tern *(Sterna forsteri)*, and Caspian Tern are also present in significant numbers. Yellow-footed Gull *(Larus livens)*, though not a breeding species, visits the sea regularly in winter from breeding grounds in the Gulf of California, another reminder of the affinity between the sea and the gulf. The Imperial Valley is also an important staging area for Black Tern in fall, which is most often seen foraging over irrigated agricultural land.

The sea is an important stopover site for migrating shorebirds as well as a wintering ground of hemispheric importance; at last count, 44 species of shorebirds had been recorded. The Pacific Flyway Project identified the Salton Sea as one of the most important sites in the interior of North America for migratory and wintering shorebirds. Western Sandpiper *(Calidris mauri)*, American Avocet *(Recurvirostra americana)*, dowitchers, Black-necked Stilt *(Himantopus mexicanus)*, Red-necked Phalarope *(Phalaropus lobatus)*, and Wilson' s Phalarope *(P. fulicarius)* are the most common species in fall passage. In winter, Willet *(Catoptrophorus semipalmatus)*, Marbled Godwit *(Limosa fedoa)*, and Least Sandpiper *(Calidris minutilla)* are numerous (tens of thousands), rivaling the Central Valley as an important interior wintering site. The sea is also the only place where Stilt Sandpiper *(C. himantopus)* overwinters in the west. Then, during spring migration Western Sandpiper, dowitchers, American Avocet, Black-necked Stilt and Whimbrel *(Numenius phaeopus)* become more numerous, passing northward toward their breeding grounds. Breeding shorebirds are restricted to Snowy Plover *(Charadrius alexandrinus)*, Killdeer *(Charadrius vociferus)*, Black-necked Stilt, and American Avocet.

Rare breeders with restricted distribution around the sea and in the valley include Least Bittern *(Ixobrychus exilis)*, California Black Rail *(Laterallus jamaicensis coturniculus)*, and Yuma Clapper

Plate 138. A Least Bittern hunts by stalking surreptitiously along marshland edges. It is a rarely seen resident in wetlands of the Central and Owens Valleys, and the Klamath area.

Rail *(Rallus longirostris yumanensis)*; unconfirmed but suspected nesting species include American Bittern *(Botaurus lentiginosus)*, Tricolored Heron *(Egretta tricolor)*, Sora *(Porzana carolina)*, and Inca Dove *(Columbina inca)*. Former breeders within the Salton Sink (or along the Colorado River) now apparently extirpated or nearly so are Wood Stork *(Mycteria americana)*, Fulvous Whistling-Duck, Harris's ("Bay-winged") Hawk *(Parabuteo unicinctus)*, Yellow-billed Cuckoo *(Coccyzus americanus)*, Elf Owl, Gilded Flicker *(Colaptes chrysoides)*, Vermilion Flycatcher *(Pyrocephalus rubinus)*, Willow Flycatcher *(Empidonax traillii)*, Bell's Vireo *(Vireo belii arizonae)*, and Lucy's Warbler.

The proportion of vagrants on the Salton Sea's bird list rival any hot spot in California, or perhaps anywhere in continental North America; approximately one-third of the species on the list are "extremely rare" in every season that they occur. As with most large bodies of landlocked waters, pelagic vagrants visit in passage; oceanic species such as loons, jaegers, Black-legged Kittiwake *(Rissa tridactyla)*, and Arctic Tern *(Sterna paradisaea)* have come to be anticipated, but are always a surprise. The list of pelagics from the sea is impressive indeed, including eight species of tubenoses, a group of birds that are highly pelagic. Where else

Plate 139. The American Bittern is a common bird of wetlands, heard more often than seen. Its cryptic plumage enables it to blend into the dun-brown grasses, cattails, and sedges of its habitat.

in the world might a Laysan Albatross *(Phoebastria immutabilis)* hear the song of a Crissal Thrasher? And most of the vagrant warblers that have occurred in California have occurred likewise at the sea.

Colorado River Valley

When naturalist Aldo Leopold visited the lower Colorado River in 1922, he found extensive forests of cottonwoods, willows, and mesquite interspersed with marshes and lagoons. Most of this wilderness has been subsequently converted, through water diversions, drainage, and tillage, to agricultural fields and salt flats. Adding insult to injury, tamarisk has colonized and dominated the riparian habitat and altered the nature of the river valley inexorably. Rosenberg et al. (1991) describe the changes in the flora and fauna in detail in their excellent *Birds of the Lower Colorado Valley.* They point out that the most dramatic declines have been suffered by those species that were dependent on the multistoried cottonwood–willow forests that once proliferated in the floodplain. Summer residents that were once common—Yellow-billed Cuckoo, Willow and Vermilion Flycatchers, Bell's Vireo, Yellow Warbler *(Dendroica petechia)*, Yellow-breasted Chat, and Summer Tanager—are now rare. Likewise, cavity-nesters that depended on the soft wood and snags of cottonwoods, Gilded Flicker and Gila Woodpecker, have declined, or, like the Elf Owl, all but disappeared. The loss of nesting opportunities has been exacerbated by competition for available cavity nest sites with the aggressive European Starling *(Sturnus vulgaris)*, a species whose increase since the mid-1950s has been as dramatic as the decline of other native species. Brown-headed Cowbird *(Molothrus ater)* and Bronzed Cowbird *(Molothrus aeneus)* and Great-tailed Grackles have also increased concurrently with the reduction in riparian vegetation, a situation that has not served the native avifauna well.

Some species, especially those that are mesquite associates—Verdin, Black-tailed Gnatcatcher, Lucy's Warbler, and Bullock's Oriole—have weathered the habitat changes better, and a few—Blue Grosbeak *(Passerina caerulea)*, Mourning and White-winged Doves—have adapted to the new tamarisk habitat and increased accordingly. Some species may have responded to, or at least have not been negatively affected by, an increase in the extent of edge habitat that has been created with agricultural developments alongside watercourses.

The impoundment of water has had a devastating effect on riparian birds, but has attracted some waterbirds that were formerly absent and increased the numbers of others that were formerly uncommon or scarce. One striking example is the Yuma Clapper Rail, a federally endangered marsh bird that has colonized the marshes associated with the impoundments during the latter half of the twentieth century. Agricultural activity, too, has eliminated habitat for some species and improved it for others:

> Again we are witnessing an ecological trade-off, with declines of native, riparian-breeding birds being offset by the establishment and expansion of species associated with open-water, marsh, and agricultural habitats. . . . [Nevertheless] immediate action needs to be taken to preserve the valley's original avifauna associated with pristine riparian habitats. (Rosenberg et al. 1991)

The recent history of one species in California—Brown-crested Flycatcher—illustrates how difficult it is to make assumptions about direct relationships between habitat and distribution. This flycatcher is found primarily in mature cottonwood–willow forests, but its population has increased in California coincident with the elimination of much of this habitat! It was not recorded breeding in California until the late 1940s, but subsequently has become well established in the southeastern-most corner of the State. This expansion of its range from south to north in the latter half of the twentieth century, though seemingly anomalous, may be related to small but progressive increases in rainfall and temperature over the same period (Johnson 1994).

In spite of all the changes induced by human endeavors, and the inevitable ecological shifts caused by diversion and modification of a naturally dynamic hydrology, the lower Colorado River valley remains an important and vital habitat for California's birdlife. The year-round residents are those expected in the Colorado deserts—Gambel's Quail, Mourning Dove, Greater Roadrunner, Gila and Ladder-backed Woodpeckers, Verdin, Black-tailed Gnatcatcher, Crissal Thrasher, Abert's Towhee, and House

Plate 140. The Cattle Egret, a fairly recent immigrant from South America, is now a well-established resident in the irrigated southern regions of the state. It feeds on insects disturbed by the movement of grazing livestock.

Plate 141. The courtship ritual of Great Egrets reinforces pair bonding. The elaborate plumes are developed and displayed in the breeding season.

Finch are all common, though in lesser numbers than under former pristine conditions. The development of impounded water bodies now supports numbers of Pied-billed Grebe, Double-crested Cormorant, Great Blue Heron, Great Egret *(Ardea alba)*, Snowy Egret *(Egretta thula)*, Black-crowned Night Heron, and American Coot *(Fulica americana)*.

Large numbers of birds, about evenly split between landbird and aquatic species, arrive to spend the winter months in the lower Colorado River valley, and the river and adjacent agricultural lands provide important winter habitat for several species of conservation concern, among them—Sandhill Crane *(Grus canadensis)*, Northern Harrier *(Circus cyaneus)*, Sharp-shinned and Cooper's Hawks *(Accipiter striatus* and *A. cooperii)*, and Loggerhead Shrike. Waterfowl and shorebirds are not particularly well represented in winter, but small flocks of geese and dabbling and diving ducks frequent Imperial National Wildlife Refuge north of Winterhaven, California, and Lake Havasu National Wildlife Refuge across the river from Needles, in Arizona. West Pond, above the Imperial Dam on the west side of the river, supports reliable marsh habitat for Black Rail *(Laterallus jamaicensis)* and Clapper Rail *(Rallus longirostris)*, one of only a few locations in southeastern California.

Migration through the lower Colorado is a lively and protracted phenomenon, and Rosenberg et al. (1991) estimate that 70 percent of the 440 or so species of migratory North American birds have been found in the valley. Spring migration begins as early as January with the arrival of hummingbirds and swallows and extends into June, when flycatchers and warblers are still passing through. No sooner is spring migration winding down than the first pulses of fall migration begin with the arrival of waterbirds—ibis, pelicans, stilts, and terns—followed by ducks, swallows, and other landbirds. As with other southern California riparian areas, passerine migration extends well into October. Accessible riparian patches are sporadically distributed from about 12 miles north of Needles (near the point where the Nevada, California, and Arizona state lines meet) southward through San Bernardino and Riverside Counties to Laguna Dam in Imperial County. The diminution of passerine migration coincides (coincidentally) with the peak arrivals of some of the most common autumn aquatic birds—American White Pelican,

Plate 142. The male Vermilion Flycatcher provides a brilliant flash of color in the desert landscape. It is a small, active bird, moving from one perch to the next, from which it darts out to capture flying insects.

White-faced Ibis, and American Avocet—and the arrival of wintering raptors.

The lower Colorado River valley, and the changes and abuses it has suffered at the hands of nature's most prodigious species, exemplified most profoundly by the diminution of its riparian forests, is a microcosm for the rest of California. It may still be possible to recover and restore a small portion of what has been lost, but this will require even more prodigious efforts and the common vision of us all. It will require the wise management of public lands, the cooperation of private land owners and public agencies, and informed restoration plans backed by reliable federal funds. The nineteenth century was one of exploration and conquest of the west; the twentieth century was a time of development and extraction of natural resources. The twenty-first century may prove to be one of restoration and rehabilitation, a period in which we, as a species dependent on the same ecosystems and natural processes as other species, recognize our responsibility and the mutual benefit we will enjoy from the preservation and the stewardship of nature.

CHECKLIST OF CALIFORNIA BIRDS

This list of 624 bird species documented in the state is adapted from the official list maintained by the Western Field Ornithologists' California Bird Records Committee (www.wfo-cbrc.org). The sequence of the 624 species of birds on this list follows the seventh edition of the American Ornithologists Union Check-list of North American Birds (1998) and supplements, as adapted from the Western Field Ornithologists' California Bird Records Committee list (July 30, 2004), www.wfo-cbrc.org/cbrc/ca_list.html.

I Introduced; nonnative but now established in California (10 species)
FE Federally endangered
SE State endangered
FT Federally threatened
ST State threatened
E Extirpated from California (two species)
X Extinct

Screamers, Ducks, Geese, and Swans (Anseriformes)

DUCKS, GEESE, AND SWANS (ANATIDAE)

- ☐ Black-bellied Whistling-Duck *(Dendrocygna autumnalis)*
- ☐ Fulvous Whistling-Duck *(Dendrocygna bicolor)*
- ☐ Greater White-fronted Goose *(Anser albifrons)*
- ☐ Emperor Goose *(Chen canagica)*
- ☐ Snow Goose *(Chen caerulescens)*
- ☐ Ross's Goose *(Chen rossii)*
- ☐ Cackling Goose *(Branta hutchinsii)*
- ☐ Canada Goose *(Branta canadensis)*
- ☐ Brant *(Branta bernicla)*

- ☐ Trumpeter Swan *(Cygnus buccinator)*
- ☐ Tundra Swan *(Cygnus columbianus)*
- ☐ Whooper Swan *(Cygnus cygnus)*
- ☐ Wood Duck *(Aix sponsa)*
- ☐ Gadwall *(Anas strepera)*
- ☐ Falcated Duck *(Anas falcata)*
- ☐ Eurasian Wigeon *(Anas penelope)*
- ☐ American Wigeon *(Anas americana)*
- ☐ American Black Duck *(Anas rubripes)*
- ☐ Mallard *(Anas platyrhynchos)*
- ☐ Blue-winged Teal *(Anas discors)*
- ☐ Cinnamon Teal *(Anas cyanoptera)*
- ☐ Northern Shoveler *(Anas clypeata)*
- ☐ Northern Pintail *(Anas acuta)*
- ☐ Garganey *(Anas querquedula)*
- ☐ Baikal Teal *(Anas formosa)*
- ☐ Green-winged Teal *(Anas crecca)*
- ☐ Canvasback *(Aythya valisineria)*
- ☐ Redhead *(Aythya americana)*
- ☐ Common Pochard *(Aythya ferina)*
- ☐ Ring-necked Duck *(Aythya collaris)*
- ☐ Tufted Duck *(Aythya fuligula)*
- ☐ Greater Scaup *(Aythya marila)*
- ☐ Lesser Scaup *(Aythya affinis)*
- ☐ Steller's Eider *(Polysticta stelleri)*
- ☐ King Eider *(Somateria spectabilis)*
- ☐ Harlequin Duck *(Histrionicus histrionicus)*
- ☐ Surf Scoter *(Melanitta perspicillata)*
- ☐ White-winged Scoter *(Melanitta fusca)*
- ☐ Black Scoter *(Melanitta nigra)*
- ☐ Long-tailed Duck *(Clangula hyemalis)*
- ☐ Bufflehead *(Bucephala albeola)*
- ☐ Common Goldeneye *(Bucephala clangula)*
- ☐ Barrow's Goldeneye *(Bucephala islandica)*
- ☐ Smew *(Mergellus albellus)*
- ☐ Hooded Merganser *(Lophodytes cucullatus)*
- ☐ Common Merganser *(Mergus merganser)*
- ☐ Red-breasted Merganser *(Mergus serrator)*
- ☐ Ruddy Duck *(Oxyura jamaicensis)*

Gallinaceous Birds (Galliformes)

PARTRIDGES, GROUSE, TURKEYS, AND OLD WORLD QUAIL (PHASIANIDAE)

- ☐ Chukar *(Alectoris chukar)*, I
- ☐ Ring-necked Pheasant *(Phasianus colchicus)*, I
- ☐ Ruffed Grouse *(Bonasa umbellus)*
- ☐ Greater Sage-Grouse *(Centrocercus urophasianus)*
- ☐ White-tailed Ptarmigan *(Lagopus leucurus)*, I
- ☐ Blue Grouse *(Dendragapus obscurus)*
- ☐ Sharp-tailed Grouse *(Tympanuchus phasianellus)*, E
- ☐ Wild Turkey *(Meleagris gallopavo)*, I

NEW WORLD QUAIL (ODONTOPHORIDAE)

- ☐ Mountain Quail *(Oreortyx pictus)*
- ☐ California Quail *(Callipepla californica)*
- ☐ Gambel's Quail *(Callipepla gambelii)*

Loons (Gaviiformes)

LOONS (GAVIIDAE)

- ☐ Red-throated Loon *(Gavia stellata)*
- ☐ Arctic Loon *(Gavia arctica)*
- ☐ Pacific Loon *(Gavia pacifica)*
- ☐ Common Loon *(Gavia immer)*
- ☐ Yellow-billed Loon *(Gavia adamsii)*

Grebes (Podicipediformes)

GREBES (PODICIPEDIDAE)

- ☐ Least Grebe *(Tachybaptus dominicus)*
- ☐ Pied-billed Grebe *(Podilymbus podiceps)*
- ☐ Horned Grebe *(Podiceps auritus)*
- ☐ Red-necked Grebe *(Podiceps grisegena)*
- ☐ Eared Grebe *(Podiceps nigricollis)*
- ☐ Western Grebe *(Aechmophorus occidentalis)*
- ☐ Clark's Grebe *(Aechmophorus clarkii)*

Tube-nosed Swimmers (Procellariiformes)

ALBATROSSES (DIOMEDEIDAE)

- ☐ Shy Albatross *(Thalassarche cauta)*
- ☐ Light-mantled Albatross *(Phoebetria palpebrata)*
- ☐ Wandering Albatross *(Diomedea exulans)*
- ☐ Laysan Albatross *(Phoebastria immutabilis)*

- ☐ Black-footed Albatross *(Phoebastria nigripes)*
- ☐ Short-tailed Albatross *(Phoebastria albatrus)*, FE

SHEARWATERS AND PETRELS (PROCELLARIIDAE)

- ☐ Northern Fulmar *(Fulmarus glacialis)*
- ☐ Great-winged Petrel *(Pterodroma macroptera)*
- ☐ Murphy's Petrel *(Pterodroma ultima)*
- ☐ Mottled Petrel *(Pterodroma inexpectata)*
- ☐ Dark-rumped Petrel *(Pterodroma phaeopygia/sandwichensis)*
- ☐ Cook's Petrel *(Pterodroma cookii)*
- ☐ Stejneger's Petrel *(Pterodroma longirostris)*
- ☐ Bulwer's Petrel *(Bulweria bulwerii)*
- ☐ Streaked Shearwater *(Calonectris leucomelas)*
- ☐ Cory's Shearwater *(Calonectris diomedea)*
- ☐ Pink-footed Shearwater *(Puffinus creatopus)*
- ☐ Flesh-footed Shearwater *(Puffinus carneipes)*
- ☐ Greater Shearwater *(Puffinus gravis)*
- ☐ Wedge-tailed Shearwater *(Puffinus pacificus)*
- ☐ Buller's Shearwater *(Puffinus bulleri)*
- ☐ Sooty Shearwater *(Puffinus griseus)*
- ☐ Short-tailed Shearwater *(Puffinus tenuirostris)*
- ☐ Manx Shearwater *(Puffinus puffinus)*
- ☐ Black-vented Shearwater *(Puffinus opisthomelas)*

STORM-PETRELS (HYDROBATIDAE)

- ☐ Wilson's Storm-Petrel *(Oceanites oceanicus)*
- ☐ Fork-tailed Storm-Petrel *(Oceanodroma furcata)*
- ☐ Leach's Storm-Petrel *(Oceanodroma leucorhoa)*
- ☐ Ashy Storm-Petrel *(Oceanodroma homochroa)*
- ☐ Wedge-rumped Storm-Petrel *(Oceanodroma tethys)*
- ☐ Black Storm-Petrel *(Oceanodroma melania)*
- ☐ Least Storm-Petrel *(Oceanodroma microsoma)*

Totipalmate Birds (Pelecaniformes)

TROPICBIRDS (PHAETHONTIDAE)

- ☐ White-tailed Tropicbird *(Phaethon lepturus)*
- ☐ Red-billed Tropicbird *(Phaethon aethereus)*
- ☐ Red-tailed Tropicbird *(Phaethon rubricauda)*

BOOBIES AND GANNETS (SULIDAE)

- ☐ Masked Booby *(Sula dactylatra)*
- ☐ Blue-footed Booby *(Sula nebouxii)*

- ☐ Brown Booby *(Sula leucogaster)*
- ☐ Red-footed Booby *(Sula sula)*

PELICANS (PELECANIDAE)

- ☐ American White Pelican *(Pelecanus erythrorhynchos)*
- ☐ Brown Pelican *(Pelecanus occidentalis)*, SE, FE *(californicus)*

CORMORANTS (PHALACROCORACIDAE)

- ☐ Brandt's Cormorant *(Phalacrocorax penicillatus)*
- ☐ Neotropic Cormorant *(Phalacrocorax brasilianus)*
- ☐ Double-crested Cormorant *(Phalacrocorax auritus)*
- ☐ Pelagic Cormorant *(Phalacrocorax pelagicus)*

DARTERS (ANHINGIDAE)

- ☐ Anhinga *(Anhinga anhinga)*

FRIGATEBIRDS (FREGATIDAE)

- ☐ Magnificent Frigatebird *(Fregata magnificens)*
- ☐ Great Frigatebird *(Fregata minor)*

Herons, Ibises, Storks, American Vultures, and Allies (Ciconiiformes)

HERONS, BITTERNS, AND ALLIES (ARDEIDAE)

- ☐ American Bittern *(Botaurus lentiginosus)*
- ☐ Least Bittern *(Ixobrychus exilis)*
- ☐ Great Blue Heron *(Ardea herodias)*
- ☐ Great Egret *(Ardea alba)*
- ☐ Snowy Egret *(Egretta thula)*
- ☐ Little Blue Heron *(Egretta caerulea)*
- ☐ Tricolored Heron *(Egretta tricolor)*
- ☐ Reddish Egret *(Egretta rufescens)*
- ☐ Cattle Egret *(Bubulcus ibis)*
- ☐ Green Heron *(Butorides virescens)*
- ☐ Black-crowned Night-Heron *(Nycticorax nycticorax)*
- ☐ Yellow-crowned Night-Heron *(Nyctanassa violacea)*

IBISES AND SPOONBILLS (THRESKIORNITHIDAE)

- ☐ White Ibis *(Eudocimus albus)*
- ☐ Glossy Ibis *(Plegadis falcinellus)*
- ☐ White-faced Ibis *(Plegadis chihi)*
- ☐ Roseate Spoonbill *(Platalea ajaja)*

STORKS (CICONIIDAE)

- ☐ Wood Stork *(Mycteria americana)*

NEW WORLD VULTURES (CATHARTIDAE)

- ☐ Black Vulture *(Coragyps atratus)*
- ☐ Turkey Vulture *(Cathartes aura)*
- ☐ California Condor *(Gymnogyps californianus)*, E, SE, FE

Diurnal Birds of Prey (Falconiformes)

HAWKS, KITES, EAGLES, AND ALLIES (ACCIPITRIDAE)

- ☐ Osprey *(Pandion haliaetus)*
- ☐ White-tailed Kite *(Elanus leucurus)*
- ☐ Mississippi Kite *(Ictinia mississippiensis)*
- ☐ Bald Eagle *(Haliaeetus leucocephalus)*, SE, FT
- ☐ Northern Harrier *(Circus cyaneus)*
- ☐ Sharp-shinned Hawk *(Accipiter striatus)*
- ☐ Cooper's Hawk *(Accipiter cooperii)*
- ☐ Northern Goshawk *(Accipiter gentilis)*
- ☐ Common Black-Hawk *(Buteogallus anthracinus)*
- ☐ Harris's Hawk *(Parabuteo unicinctus)*
- ☐ Red-shouldered Hawk *(Buteo lineatus)*
- ☐ Broad-winged Hawk *(Buteo platypterus)*
- ☐ Swainson's Hawk *(Buteo swainsoni)*, ST
- ☐ Zone-tailed Hawk *(Buteo albonotatus)*
- ☐ Red-tailed Hawk *(Buteo jamaicensis)*
- ☐ Ferruginous Hawk *(Buteo regalis)*
- ☐ Rough-legged Hawk *(Buteo lagopus)*
- ☐ Golden Eagle *(Aquila chrysaetos)*

CARACARAS AND FALCONS (FALCONIDAE)

- ☐ American Kestrel *(Falco sparverius)*
- ☐ Merlin *(Falco columbarius)*
- ☐ Gyrfalcon *(Falco rusticolus)*
- ☐ Peregrine Falcon *(Falco peregrinus)*
- ☐ Prairie Falcon *(Falco mexicanus)*

Rails, Cranes, and Allies (Gruiformes)

RAILS, GALLINULES, AND COOTS (RALLIDAE)

- ☐ Yellow Rail *(Coturnicops noveboracensis)*
- ☐ Black Rail *(Laterallus jamaicensis)*, ST *(cortuniculus)*
- ☐ Clapper Rail *(Rallus longirostris)*, SE, FE *(obsoletus)*, *(levipes)*; SE *(yumanensis)*
- ☐ Virginia Rail *(Rallus limicola)*
- ☐ Sora *(Porzana carolina)*

- ☐ Purple Gallinule *(Porphyrio martinica)*
- ☐ Common Moorhen *(Gallinula chloropus)*
- ☐ American Coot *(Fulica americana)*

CRANES (GRUIDAE)

- ☐ Sandhill Crane *(Grus canadensis)*, ST *(tabida)*

Shorebirds, Gulls, Auks, and Allies (Charadriiformes)

LAPWINGS AND PLOVERS (CHARADRIIDAE)

- ☐ Black-bellied Plover *(Pluvialis squatarola)*
- ☐ American Golden-Plover *(Pluvialis dominica)*
- ☐ Pacific Golden-Plover *(Pluvialis fulva)*
- ☐ Mongolian Plover *(Charadrius mongolus)*
- ☐ Greater Sandplover *(Charadrius leschenaultii)*
- ☐ Snowy Plover *(Charadrius alexandrinus)*, FT *(nivosus)*
- ☐ Wilson's Plover *(Charadrius wilsonia)*
- ☐ Semipalmated Plover *(Charadrius semipalmatus)*
- ☐ Piping Plover *(Charadrius melodus)*
- ☐ Killdeer *(Charadrius vociferus)*
- ☐ Mountain Plover *(Charadrius montanus)*
- ☐ Eurasian Dotterel *(Charadrius morinellus)*

OYSTERCATCHERS (HAEMATOPODIDAE)

- ☐ American Oystercatcher *(Haematopus palliatus)*
- ☐ Black Oystercatcher *(Haematopus bachmani)*

STILTS AND AVOCETS (RECURVIROSTRIDAE)

- ☐ Black-necked Stilt *(Himantopus mexicanus)*
- ☐ American Avocet *(Recurvirostra americana)*

SANDPIPERS, PHALAROPES, AND ALLIES (SCOLOPACIDAE)

- ☐ Common Greenshank *(Tringa nebularia)*
- ☐ Greater Yellowlegs *(Tringa melanoleuca)*
- ☐ Lesser Yellowlegs *(Tringa flavipes)*
- ☐ Spotted Redshank *(Tringa erythropus)*
- ☐ Solitary Sandpiper *(Tringa solitaria)*
- ☐ Willet *(Catoptrophorus semipalmatus)*
- ☐ Wandering Tattler *(Heteroscelus incanus)*
- ☐ Gray-tailed Tattler *(Heteroscelus brevipes)*
- ☐ Spotted Sandpiper *(Actitis macularius)*
- ☐ Terek Sandpiper *(Xenus cinereus)*
- ☐ Upland Sandpiper *(Bartramia longicauda)*

- ☐ Little Curlew *(Numenius minutus)*
- ☐ Whimbrel *(Numenius phaeopus)*
- ☐ Bristle-thighed Curlew *(Numenius tahitiensis)*
- ☐ Long-billed Curlew *(Numenius americanus)*
- ☐ Hudsonian Godwit *(Limosa haemastica)*
- ☐ Bar-tailed Godwit *(Limosa lapponica)*
- ☐ Marbled Godwit *(Limosa fedoa)*
- ☐ Ruddy Turnstone *(Arenaria interpres)*
- ☐ Black Turnstone *(Arenaria melanocephala)*
- ☐ Surfbird *(Aphriza virgata)*
- ☐ Red Knot *(Calidris canutus)*
- ☐ Sanderling *(Calidris alba)*
- ☐ Semipalmated Sandpiper *(Calidris pusilla)*
- ☐ Western Sandpiper *(Calidris mauri)*
- ☐ Red-necked Stint *(Calidris ruficollis)*
- ☐ Little Stint *(Calidris minuta)*
- ☐ Long-toed Stint *(Calidris subminuta)*
- ☐ Least Sandpiper *(Calidris minutilla)*
- ☐ White-rumped Sandpiper *(Calidris fuscicollis)*
- ☐ Baird's Sandpiper *(Calidris bairdii)*
- ☐ Pectoral Sandpiper *(Calidris melanotos)*
- ☐ Sharp-tailed Sandpiper *(Calidris acuminata)*
- ☐ Rock Sandpiper *(Calidris ptilocnemis)*
- ☐ Dunlin *(Calidris alpina)*
- ☐ Curlew Sandpiper *(Calidris ferruginea)*
- ☐ Stilt Sandpiper *(Calidris himantopus)*
- ☐ Buff-breasted Sandpiper *(Tryngites subruficollis)*
- ☐ Ruff *(Philomachus pugnax)*
- ☐ Short-billed Dowitcher *(Limnodromus griseus)*
- ☐ Long-billed Dowitcher *(Limnodromus scolopaceus)*
- ☐ Jack Snipe *(Lymnocryptes minimus)*
- ☐ Wilson's Snipe *(Gallinago delicata)*
- ☐ American Woodcock *(Scolopax minor)*
- ☐ Wilson's Phalarope *(Phalaropus tricolor)*
- ☐ Red-necked Phalarope *(Phalaropus lobatus)*
- ☐ Red Phalarope *(Phalaropus fulicarius)*

SKUAS, GULLS, TERNS, AND SKIMMERS (LARIDAE)

- ☐ South Polar Skua *(Catharacta maccormicki)*
- ☐ Pomarine Jaeger *(Stercorarius pomarinus)*

☐ Parasitic Jaeger *(Stercorarius parasiticus)*
☐ Long-tailed Jaeger *(Stercorarius longicaudus)*
☐ Laughing Gull *(Larus atricilla)*
☐ Franklin's Gull *(Larus pipixcan)*
☐ Little Gull *(Larus minutus)*
☐ Black-headed Gull *(Larus ridibundus)*
☐ Bonaparte's Gull *(Larus philadelphia)*
☐ Heermann's Gull *(Larus heermanni)*
☐ Belcher's Gull *(Larus belcheri)*
☐ Black-tailed Gull *(Larus crassirostris)*
☐ Mew Gull *(Larus canus)*
☐ Ring-billed Gull *(Larus delawarensis)*
☐ California Gull *(Larus californicus)*
☐ Herring Gull *(Larus argentatus)*
☐ Thayer's Gull *(Larus thayeri)*
☐ Iceland Gull *(Larus glaucoides)*
☐ Lesser Black-backed Gull *(Larus fuscus)*
☐ Yellow-footed Gull *(Larus livens)*
☐ Western Gull *(Larus occidentalis)*
☐ Glaucous-winged Gull *(Larus glaucescens)*
☐ Glaucous Gull *(Larus hyperboreus)*
☐ Sabine's Gull *(Xema sabini)*
☐ Swallow-tailed Gull *(Creagrus furcatus)*
☐ Black-legged Kittiwake *(Rissa tridactyla)*
☐ Red-legged Kittiwake *(Rissa brevirostris)*
☐ Ivory Gull *(Pagophila eburnea)*
☐ Gull-billed Tern *(Sterna nilotica)*
☐ Caspian Tern *(Sterna caspia)*
☐ Royal Tern *(Sterna maxima)*
☐ Elegant Tern *(Sterna elegans)*
☐ Sandwich Tern *(Sterna)*
☐ Common Tern *(Sterna hirundo)*
☐ Arctic Tern *(Sterna paradisaea)*
☐ Forster's Tern *(Sterna forsteri)*
☐ Least Tern *(Sterna antillarum)*
☐ Bridled Tern *(Sterna anaethetus)*
☐ Sooty Tern *(Sterna fuscata)*
☐ White-winged Tern *(Chlidonias leucopterus)*
☐ Black Tern *(Chlidonias niger)*
☐ Black Skimmer *(Rynchops niger)*

AUKS, MURRES, AND PUFFINS (ALCIDAE)

- ☐ Common Murre *(Uria aalge)*
- ☐ Thick-billed Murre *(Uria lomvia)*
- ☐ Pigeon Guillemot *(Cepphus columba)*
- ☐ Long-billed Murrelet *(Brachyramphus perdix)*
- ☐ Marbled Murrelet *(Brachyramphus marmoratus)*, SE, FT
- ☐ Kittlitz's Murrelet *(Brachyramphus brevirostris)*
- ☐ Xantus's Murrelet *(Synthliboramphus hypoleucus)*
- ☐ Craveri's Murrelet *(Synthliboramphus craveri)*
- ☐ Ancient Murrelet *(Synthliboramphus antiquus)*
- ☐ Cassin's Auklet *(Ptychoramphus aleuticus)*
- ☐ Parakeet Auklet *(Aethia psittacula)*
- ☐ Least Auklet *(Aethia pusilla)*
- ☐ Crested Auklet *(Aethia cristatella)*
- ☐ Rhinoceros Auklet *(Cerorhinca monocerata)*
- ☐ Horned Puffin *(Fratercula corniculata)*
- ☐ Tufted Puffin *(Fratercula cirrhata)*

Pigeons and Doves (Columbiformes)

PIGEONS AND DOVES (COLUMBIDAE)

- ☐ Rock Pigeon (formerly Rock Dove) *(Columba livia)*, I
- ☐ Band-tailed Pigeon *(Patagioenas fasciata)*
- ☐ Eurasian Collared-Dove *(Streptopelia decaocto)*, I
- ☐ Spotted Dove *(Streptopelia chinensis)*, I
- ☐ White-winged Dove *(Zenaida asiatica)*
- ☐ Mourning Dove *(Zenaida macroura)*
- ☐ Inca Dove *(Columbina inca)*
- ☐ Common Ground-Dove *(Columbina passerina)*
- ☐ Ruddy Ground-Dove *(Columbina talpacoti)*

Parrots (Psittaciformes)

LORIES, PARAKEETS, MACAWS, AND PARROTS (PSITTACIDAE)

- ☐ Red-crowned Parrot *(Amazona viridigenalis)*, I

Cuckoos and Allies (Cuculiformes)

CUCKOOS, ROADRUNNERS, AND ANIS (CUCULIDAE)

- ☐ Black-billed Cuckoo *(Coccyzus erythropthalmus)*
- ☐ Yellow-billed Cuckoo *(Coccyzus americanus)*
- ☐ Greater Roadrunner *(Geococcyx californianus)*
- ☐ Groove-billed Ani *(Crotophaga sulcirostris)*

Owls (Strigiformes)

BARN OWLS (TYTONIDAE)

- ☐ Barn Owl *(Tyto alba)*

TYPICAL OWLS (STRIGIDAE)

- ☐ Flammulated Owl *(Otus flammeolus)*
- ☐ Western Screech-Owl *(Megascops kennicottii)*
- ☐ Great Horned Owl *(Bubo virginianus)*
- ☐ Snowy Owl *(Bubo scandiacus)*
- ☐ Northern Pygmy-Owl *(Glaucidium gnoma)*
- ☐ Elf Owl *(Micrathene whitneyi)*
- ☐ Burrowing Owl *(Athene cunicularia)*
- ☐ Spotted Owl *(Strix occidentalis)*, FT *(caurina)*
- ☐ Barred Owl *(Strix varia)*
- ☐ Great Gray Owl *(Strix nebulosa)*, SE
- ☐ Long-eared Owl *(Asio otus)*
- ☐ Short-eared Owl *(Asio flammeus)*
- ☐ Northern Saw-whet Owl *(Aegolius acadicus)*

Goatsuckers, Oilbirds, and Allies (Caprimulgiformes)

GOATSUCKERS (CAPRIMULGIDAE)

- ☐ Lesser Nighthawk *(Chordeiles acutipennis)*
- ☐ Common Nighthawk *(Chordeiles minor)*
- ☐ Common Poorwill *(Phalaenoptilus nuttallii)*
- ☐ Chuck-will's-widow *(Caprimulgus carolinensis)*
- ☐ Buff-collared Nightjar *(Caprimulgus ridgwayi)*
- ☐ Whip-poor-will *(Caprimulgus vociferus)*

Swifts and Hummingbirds (Apodiformes)

SWIFTS (APODIDAE)

- ☐ Black Swift *(Cypseloides niger)*
- ☐ White-collared Swift *(Streptoprocne zonaris)*
- ☐ Chimney Swift *(Chaetura pelagica)*
- ☐ Vaux's Swift *(Chaetura vauxi)*
- ☐ White-throated Swift *(Aeronautes saxatalis)*

HUMMINGBIRDS (TROCHILIDAE)

- ☐ Broad-billed Hummingbird *(Cynanthus latirostris)*
- ☐ Xantus's Hummingbird *(Hylocharis xantusii)*
- ☐ Violet-crowned Hummingbird *(Amazilia violiceps)*
- ☐ Blue-throated Hummingbird *(Lampornis clemenciae)*

- ☐ Magnificent Hummingbird *(Eugenes fulgens)*
- ☐ Ruby-throated Hummingbird *(Archilochus colubris)*
- ☐ Black-chinned Hummingbird *(Archilochus alexandri)*
- ☐ Anna's Hummingbird *(Calypte anna)*
- ☐ Costa's Hummingbird *(Calypte costae)*
- ☐ Calliope Hummingbird *(Stellula calliope)*
- ☐ Broad-tailed Hummingbird *(Selasphorus platycercus)*
- ☐ Rufous Hummingbird *(Selasphorus rufus)*
- ☐ Allen's Hummingbird *(Selasphorus sasin)*

Rollers, Motmots, Kingfishers, and Allies (Coraciiformes)

KINGFISHERS (ALCEDINIDAE)

- ☐ Belted Kingfisher *(Ceryle alcyon)*

Puffbirds, Jacamars, Toucans, Woodpeckers, and Allies (Piciformes)

WOODPECKERS AND ALLIES (PICIDAE)

- ☐ Lewis's Woodpecker *(Melanerpes lewis)*
- ☐ Red-headed Woodpecker *(Melanerpes erythrocephalus)*
- ☐ Acorn Woodpecker *(Melanerpes formicivorus)*
- ☐ Gila Woodpecker *(Melanerpes uropygialis)*, SE
- ☐ Williamson's Sapsucker *(Sphyrapicus thyroideus)*
- ☐ Yellow-bellied Sapsucker *(Sphyrapicus varius)*
- ☐ Red-naped Sapsucker *(Sphyrapicus nuchalis)*
- ☐ Red-breasted Sapsucker *(Sphyrapicus ruber)*
- ☐ Ladder-backed Woodpecker *(Picoides scalaris)*
- ☐ Nuttall's Woodpecker *(Picoides nuttallii)*
- ☐ Downy Woodpecker *(Picoides pubescens)*
- ☐ Hairy Woodpecker *(Picoides villosus)*
- ☐ White-headed Woodpecker *(Picoides albolarvatus)*
- ☐ Black-backed Woodpecker *(Picoides arcticus)*
- ☐ Northern Flicker *(Colaptes auratus)*
- ☐ Gilded Flicker *(Colaptes chrysoides)*, SE
- ☐ Pileated Woodpecker *(Dryocopus pileatus)*

Passerine Birds (Passeriformes)

TYRANT FLYCATCHERS (TYRANNIDAE)

- ☐ Olive-sided Flycatcher *(Contopus cooperi)*
- ☐ Greater Pewee *(Contopus pertinax)*
- ☐ Western Wood-Pewee *(Contopus sordidulus)*

☐ Eastern Wood-Pewee *(Contopus virens)*
☐ Yellow-bellied Flycatcher *(Empidonax flaviventris)*
☐ Alder Flycatcher *(Empidonax alnorum)*
☐ Willow Flycatcher *(Empidonax traillii)*, SE, FE *(extimus)*
☐ Least Flycatcher *(Empidonax minimus)*
☐ Hammond's Flycatcher *(Empidonax hammondii)*
☐ Gray Flycatcher *(Empidonax wrightii)*
☐ Dusky Flycatcher *(Empidonax oberholseri)*
☐ Pacific-slope Flycatcher *(Empidonax difficilis)*
☐ Cordilleran Flycatcher *(Empidonax occidentalis)*
☐ Black Phoebe *(Sayornis nigricans)*
☐ Eastern Phoebe *(Sayornis phoebe)*
☐ Say's Phoebe *(Sayornis saya)*
☐ Vermilion Flycatcher *(Pyrocephalus rubinus)*
☐ Dusky-capped Flycatcher *(Myiarchus tuberculifer)*
☐ Ash-throated Flycatcher *(Myiarchus cinerascens)*
☐ Nutting's Flycatcher *(Myiarchus nuttingi)*
☐ Great Crested Flycatcher *(Myiarchus crinitus)*
☐ Brown-crested Flycatcher *(Myiarchus tyrannulus)*
☐ Sulphur-bellied Flycatcher *(Myiodynastes luteiventris)*
☐ Tropical Kingbird *(Tyrannus melancholicus)*
☐ Couch's Kingbird *(Tyrannus couchii)*
☐ Cassin's Kingbird *(Tyrannus vociferans)*
☐ Thick-billed Kingbird *(Tyrannus crassirostris)*
☐ Western Kingbird *(Tyrannus verticalis)*
☐ Eastern Kingbird *(Tyrannus tyrannus)*
☐ Scissor-tailed Flycatcher *(Tyrannus forficatus)*
☐ Fork-tailed Flycatcher *(Tyrannus savana)*

SHRIKES (LANIIDAE)

☐ Brown Shrike *(Lanius cristatus)*
☐ Loggerhead Shrike *(Lanius ludovicianus)*, FE *(mearnsi)*
☐ Northern Shrike *(Lanius excubitor)*

VIREOS (VIREONIDAE)

☐ White-eyed Vireo *(Vireo griseus)*
☐ Bell's Vireo *(Vireo bellii)*, SE *(arizonae)*, SE, FE *(pusillus)*
☐ Gray Vireo *(Vireo vicinior)*
☐ Yellow-throated Vireo *(Vireo flavifrons)*
☐ Plumbeous Vireo *(Vireo plumbeus)*
☐ Cassin's Vireo *(Vireo cassinii)*

- ☐ Blue-headed Vireo *(Vireo solitarius)*
- ☐ Hutton's Vireo *(Vireo huttoni)*
- ☐ Warbling Vireo *(Vireo gilvus)*
- ☐ Philadelphia Vireo *(Vireo philadelphicus)*
- ☐ Red-eyed Vireo *(Vireo olivaceus)*
- ☐ Yellow-green Vireo *(Vireo flavoviridis)*

CROWS AND JAYS (CORVIDAE)

- ☐ Gray Jay *(Perisoreus canadensis)*
- ☐ Steller's Jay *(Cyanocitta stelleri)*
- ☐ Blue Jay *(Cyanocitta cristata)*
- ☐ Island Scrub-Jay *(Aphelocoma insularis)*
- ☐ Western Scrub-Jay *(Aphelocoma californica)*
- ☐ Pinyon Jay *(Gymnorhinus cyanocephalus)*
- ☐ Clark's Nutcracker *(Nucifraga columbiana)*
- ☐ Black-billed Magpie *(Pica pica)*
- ☐ Yellow-billed Magpie *(Pica nuttalli)*
- ☐ American Crow *(Corvus brachyrhynchos)*
- ☐ Common Raven *(Corvus corax)*

LARKS (ALAUDIDAE)

- ☐ Sky Lark *(Alauda arvensis)*
- ☐ Horned Lark *(Eremophila alpestris)*

SWALLOWS (HIRUNDINIDAE)

- ☐ Purple Martin *(Progne subis)*
- ☐ Tree Swallow *(Tachycineta bicolor)*
- ☐ Violet-green Swallow *(Tachycineta thalassina)*
- ☐ Northern Rough-winged Swallow *(Stelgidopteryx serripennis)*
- ☐ Bank Swallow *(Riparia riparia)*, ST
- ☐ Cliff Swallow *(Petrochelidon pyrrhonota)*
- ☐ Cave Swallow *(Petrochelidon fulva)*
- ☐ Barn Swallow *(Hirundo rustica)*

CHICKADEES AND TITMICE (PARIDAE)

- ☐ Black-capped Chickadee *(Poecile atricapillus)*
- ☐ Mountain Chickadee *(Poecile gambeli)*
- ☐ Chestnut-backed Chickadee *(Poecile rufescens)*
- ☐ Oak Titmouse *(Baeolophus inornatus)*
- ☐ Juniper Titmouse *(Baeolophus griseus*

PENDULINE TITS AND VERDINS (REMIZIDAE)

- ☐ Verdin *(Auriparus flaviceps)*

LONG-TAILED TITS AND BUSHTITS (AEGITHALIDAE)

☐ Bushtit *(Psaltriparus minimus)*

NUTHATCHES (SITTIDAE)

☐ Red-breasted Nuthatch *(Sitta canadensis)*
☐ White-breasted Nuthatch *(Sitta carolinensis)*
☐ Pygmy Nuthatch *(Sitta pygmaea)*

CREEPERS (CERTHIIDAE)

☐ Brown Creeper *(Certhia americana)*

WRENS (TROGLODYTIDAE)

☐ Cactus Wren *(Campylorhynchus brunneicapillus)*
☐ Rock Wren *(Salpinctes obsoletus)*
☐ Canyon Wren *(Catherpes mexicanus)*
☐ Bewick's Wren *(Thryomanes bewickii)*
☐ House Wren *(Troglodytes aedon)*
☐ Winter Wren *(Troglodytes troglodytes)*
☐ Sedge Wren *(Cistothorus platensis)*
☐ Marsh Wren *(Cistothorus palustris)*

DIPPERS (CINCLIDAE)

☐ American Dipper *(Cinclus mexicanus)*

KINGLETS (REGULIDAE)

☐ Golden-crowned Kinglet *(Regulus satrapa)*
☐ Ruby-crowned Kinglet *(Regulus calendula)*

OLD WORLD WARBLERS AND GNATCATCHERS (SYLVIIDAE)

☐ Lanceolated Warbler *(Locustella lanceolata)*
☐ Dusky Warbler *(Phylloscopus fuscatus)*
☐ Arctic Warbler *(Phylloscopus borealis)*
☐ Blue-gray Gnatcatcher *(Polioptila caerulea)*
☐ California Gnatcatcher *(Polioptila californica)*, FT *(californica)*
☐ Black-tailed Gnatcatcher *(Polioptila melanura)*

THRUSHES (TURDIDAE)

☐ Red-flanked Bluetail *(Tarsiger cyanurus)*
☐ Northern Wheatear *(Oenanthe oenanthe)*
☐ Western Bluebird *(Sialia mexicana)*
☐ Mountain Bluebird *(Sialia currucoides)*
☐ Townsend's Solitaire *(Myadestes townsendi)*
☐ Veery *(Catharus fuscescens)*
☐ Gray-cheeked Thrush *(Catharus minimus)*
☐ Swainson's Thrush *(Catharus ustulatus)*

- ☐ Hermit Thrush *(Catharus guttatus)*
- ☐ Wood Thrush *(Hylocichla mustelina)*
- ☐ Eyebrowed Thrush *(Turdus obscurus)*
- ☐ Rufous-backed Robin *(Turdus rufopalliatus)*
- ☐ American Robin *(Turdus migratorius)*
- ☐ Varied Thrush *(Ixoreus naevius)*

BABBLERS (TIMALIIDAE)

- ☐ Wrentit *(Chamaea fasciata)*

MOCKINGBIRDS AND THRASHERS (MIMIDAE)

- ☐ Gray Catbird *(Dumetella carolinensis)*
- ☐ Northern Mockingbird *(Mimus polyglottos)*
- ☐ Sage Thrasher *(Oreoscoptes montanus)*
- ☐ Brown Thrasher *(Toxostoma rufum)*
- ☐ Bendire's Thrasher *(Toxostoma bendirei)*
- ☐ Curve-billed Thrasher *(Toxostoma curvirostre)*
- ☐ California Thrasher *(Toxostoma redivivum)*
- ☐ Crissal Thrasher *(Toxostoma crissale)*
- ☐ Le Conte's Thrasher *(Toxostoma lecontei)*

STARLINGS (STURNIDAE)

- ☐ European Starling *(Sturnus vulgaris)*, I

WAGTAILS AND PIPITS (MOTACILLIDAE)

- ☐ Eastern Yellow Wagtail *(Motacilla tschutschensis)*
- ☐ Gray Wagtail *(Motacilla cinerea)*
- ☐ White Wagtail *(Motacilla alba)*
- ☐ Black-backed Wagtail *(Motacilla lugens)*
- ☐ Olive-backed Pipit *(Anthus hodgsoni)*
- ☐ Red-throated Pipit *(Anthus cervinus)*
- ☐ American Pipit *(Anthus rubescens)*
- ☐ Sprague's Pipit *(Anthus spragueii)*

WAXWINGS (BOMBYCILLIDAE)

- ☐ Bohemian Waxwing *(Bombycilla garrulus)*
- ☐ Cedar Waxwing *(Bombycilla cedrorum)*

SILKY-FLYCATCHERS (PTILOGONATIDAE)

- ☐ Phainopepla *(Phainopepla nitens)*

WOOD-WARBLERS (PARULIDAE)

- ☐ Blue-winged Warbler *(Vermivora pinus)*
- ☐ Golden-winged Warbler *(Vermivora chrysoptera)*

- ☐ Tennessee Warbler *(Vermivora peregrina)*
- ☐ Orange-crowned Warbler *(Vermivora celata)*
- ☐ Nashville Warbler *(Vermivora ruficapilla)*
- ☐ Virginia's Warbler *(Vermivora virginiae)*
- ☐ Lucy's Warbler *(Vermivora luciae)*
- ☐ Northern Parula *(Parula americana)*
- ☐ Yellow Warbler *(Dendroica petechia)*
- ☐ Chestnut-sided Warbler *(Dendroica pensylvanica)*
- ☐ Magnolia Warbler *(Dendroica magnolia)*
- ☐ Cape May Warbler *(Dendroica tigrina)*
- ☐ Black-throated Blue Warbler *(Dendroica caerulescens)*
- ☐ Yellow-rumped Warbler *(Dendroica coronata)*
- ☐ Black-throated Gray Warbler *(Dendroica nigrescens)*
- ☐ Golden-cheeked Warbler *(Dendroica chrysoparia)*
- ☐ Black-throated Green Warbler *(Dendroica virens)*
- ☐ Townsend's Warbler *(Dendroica townsendi)*
- ☐ Hermit Warbler *(Dendroica occidentalis)*
- ☐ Blackburnian Warbler *(Dendroica fusca)*
- ☐ Yellow-throated Warbler *(Dendroica dominica)*
- ☐ Grace's Warbler *(Dendroica graciae)*
- ☐ Pine Warbler *(Dendroica pinus)*
- ☐ Prairie Warbler *(Dendroica discolor)*
- ☐ Palm Warbler *(Dendroica palmarum)*
- ☐ Bay-breasted Warbler *(Dendroica castanea)*
- ☐ Blackpoll Warbler *(Dendroica striata)*
- ☐ Cerulean Warbler *(Dendroica cerulea)*
- ☐ Black-and-white Warbler *(Mniotilta varia)*
- ☐ American Redstart *(Setophaga ruticilla)*
- ☐ Prothonotary Warbler *(Protonotaria citrea)*
- ☐ Worm-eating Warbler *(Helmitheros vermivorus)*
- ☐ Ovenbird *(Seiurus aurocapillus)*
- ☐ Northern Waterthrush *(Seiurus noveboracensis)*
- ☐ Louisiana Waterthrush *(Seiurus motacilla)*
- ☐ Kentucky Warbler *(Oporornis formosus)*
- ☐ Connecticut Warbler *(Oporornis agilis)*
- ☐ Mourning Warbler *(Oporornis philadelphia)*
- ☐ MacGillivray's Warbler *(Oporornis tolmiei)*
- ☐ Common Yellowthroat *(Geothlypis trichas)*
- ☐ Hooded Warbler *(Wilsonia citrina)*

- ☐ Wilson's Warbler *(Wilsonia pusilla)*
- ☐ Canada Warbler *(Wilsonia canadensis)*
- ☐ Red-faced Warbler *(Cardellina rubrifrons)*
- ☐ Painted Redstart *(Myioborus pictus)*
- ☐ Yellow-breasted Chat *(Icteria virens)*

TANAGERS (THRAUPIDAE)

- ☐ Hepatic Tanager *(Piranga flava)*
- ☐ Summer Tanager *(Piranga rubra)*
- ☐ Scarlet Tanager *(Piranga olivacea)*
- ☐ Western Tanager *(Piranga ludoviciana)*

EMBERIZIDS (EMBERIZIDAE)

- ☐ Green-tailed Towhee *(Pipilo chlorurus)*
- ☐ Spotted Towhee *(Pipilo maculatus)*
- ☐ California Towhee *(Pipilo crissalis)*, SE *(eremophilus)*
- ☐ Abert's Towhee *(Pipilo aberti)*
- ☐ Cassin's Sparrow *(Aimophila cassinii)*
- ☐ Rufous-crowned Sparrow *(Aimophila ruficeps)*
- ☐ American Tree Sparrow *(Spizella arborea)*
- ☐ Chipping Sparrow *(Spizella passerina)*
- ☐ Clay-colored Sparrow *(Spizella pallida)*
- ☐ Brewer's Sparrow *(Spizella breweri)*
- ☐ Field Sparrow *(Spizella pusilla)*
- ☐ Black-chinned Sparrow *(Spizella atrogularis)*
- ☐ Vesper Sparrow *(Pooecetes gramineus)*
- ☐ Lark Sparrow *(Chondestes grammacus)*
- ☐ Black-throated Sparrow *(Amphispiza bilineata)*
- ☐ Sage Sparrow *(Amphispiza belli)*, FT *(clementae)*
- ☐ Lark Bunting *(Calamospiza melanocorys)*
- ☐ Savannah Sparrow *(Passerculus sandwichensis)*, SE *(beldingi)*
- ☐ Grasshopper Sparrow *(Ammodramus savannarum)*
- ☐ Baird's Sparrow *(Ammodramus bairdii)*
- ☐ Le Conte's Sparrow *(Ammodramus leconteii)*
- ☐ Nelson's Sharp-tailed Sparrow *(Ammodramus nelsoni)*
- ☐ Fox Sparrow *(Passerella iliaca)*
- ☐ Song Sparrow *(Melospiza melodia)*
- ☐ Lincoln's Sparrow *(Melospiza lincolnii)*
- ☐ Swamp Sparrow *(Melospiza georgiana)*
- ☐ White-throated Sparrow *(Zonotrichia albicollis)*
- ☐ Harris's Sparrow *(Zonotrichia querula)*

- ☐ White-crowned Sparrow *(Zonotrichia leucophrys)*
- ☐ Golden-crowned Sparrow *(Zonotrichia atricapilla)*
- ☐ Dark-eyed Junco *(Junco hyemalis)*
- ☐ McCown's Longspur *(Calcarius mccownii)*
- ☐ Lapland Longspur *(Calcarius lapponicus)*
- ☐ Smith's Longspur *(Calcarius pictus)*
- ☐ Chestnut-collared Longspur *(Calcarius ornatus)*
- ☐ Little Bunting *(Emberiza pusilla)*
- ☐ Rustic Bunting *(Emberiza rustica)*
- ☐ Snow Bunting *(Plectrophenax nivalis)*

CARDINALS, SALTATORS, AND ALLIES (CARDINALIDAE)

- ☐ Northern Cardinal *(Cardinalis cardinalis)*
- ☐ Pyrrhuloxia *(Cardinalis sinuatus)*
- ☐ Rose-breasted Grosbeak *(Pheucticus ludovicianus)*
- ☐ Black-headed Grosbeak *(Pheucticus melanocephalus)*
- ☐ Blue Grosbeak *(Passerina caerulea)*
- ☐ Lazuli Bunting *(Passerina amoena)*
- ☐ Indigo Bunting *(Passerina cyanea)*
- ☐ Varied Bunting *(Passerina versicolor)*
- ☐ Painted Bunting *(Passerina ciris)*
- ☐ Dickcissel *(Spiza americana)*

BLACKBIRDS (ICTERIDAE)

- ☐ Bobolink *(Dolichonyx oryzivorus)*
- ☐ Red-winged Blackbird *(Agelaius phoeniceus)*
- ☐ Tricolored Blackbird *(Agelaius tricolor)*
- ☐ Western Meadowlark *(Sturnella neglecta)*
- ☐ Yellow-headed Blackbird *(Xanthocephalus xanthocephalus)*
- ☐ Rusty Blackbird *(Euphagus carolinus)*
- ☐ Brewer's Blackbird *(Euphagus cyanocephalus)*
- ☐ Common Grackle *(Quiscalus quiscula)*
- ☐ Great-tailed Grackle *(Quiscalus mexicanus)*
- ☐ Bronzed Cowbird *(Molothrus aeneus)*
- ☐ Brown-headed Cowbird *(Molothrus ater)*
- ☐ Orchard Oriole *(Icterus spurius)*
- ☐ Hooded Oriole *(Icterus cucullatus)*
- ☐ Streak-backed Oriole *(Icterus pustulatus)*
- ☐ Baltimore Oriole *(Icterus galbula)*
- ☐ Bullock's Oriole *(Icterus bullockii)*
- ☐ Scott's Oriole *(Icterus parisorum)*

FRINGILLINE AND CARDUELINE FINCHES AND ALLIES (FRINGILLIDAE)

☐ Brambling *(Fringilla montifringilla)*
☐ Gray-crowned Rosy-Finch *(Leucosticte tephrocotis)*
☐ Black Rosy-Finch *(Leucosticte atrata)*
☐ Pine Grosbeak *(Pinicola enucleator)*
☐ Purple Finch *(Carpodacus purpureus)*
☐ Cassin's Finch *(Carpodacus cassinii)*
☐ House Finch *(Carpodacus mexicanus)*
☐ Red Crossbill *(Loxia curvirostra)*
☐ White-winged Crossbill *(Loxia leucoptera)*
☐ Common Redpoll *(Carduelis flammea)*
☐ Pine Siskin *(Carduelis pinus)*
☐ Lesser Goldfinch *(Carduelis psaltria)*
☐ Lawrence's Goldfinch *(Carduelis lawrencei)*
☐ American Goldfinch *(Carduelis tristis)*
☐ Evening Grosbeak *(Coccothraustes vespertinus)*

OLD WORLD SPARROWS (PASSERIDAE)

☐ House Sparrow *(Passer domesticus)*, I

GLOSSARY

Accipiter A short-winged relatively long-tailed hawk in the genus *Accipiter.*

Allopatric Occurring in separate, widely differing geographic areas.

Altricial Born naked.

Amphipods Small crustaceans that comprise the primary prey items of small shorebirds.

Ardeid A member of the Ardeidae family, the bitterns and herons.

Bajada An alluvial slope, usually in the desert.

Benthic Living in or attached to the bottom, the substrate. Many marine mollusks and worms favored by diving birds are benthic organisms.

Biomass The total mass of organisms comprising all or part of a population; mass is usually calculated by weight and provides a different value than does abundance. Therefore a thousand Greater Sage-Grouse *(Centrocercus urophasianus)* would contribute a much higher proportion to the biomass of the Great Basin community than a thousand Sage Sparrows *(Amphispiza belli).*

Brood parasitism The use of a host species to raise the young of another species (the parasite).

Buteo Any of various broad-winged, soaring hawks of the genus *Buteo.*

Chaparral A mild temperate region characterized by hot, dry summers and cool, most winters (i.e., a Mediterranean climate) and dominated by dense growth of small-leaved evergreen shrubs.

Cismontane Literally "this side of the mountains," used here to mean the coastal slope.

Commensal relationship Symbiosis in which one species derives benefit from a common food supply while the other species is not adversely affected.

Congener Belonging to the same genus; a close relative.

Conspecific Belonging to the same species.

Convergent evolution The independent evolution of structural or functional similarity among unrelated groups.

Copepods Tiny crustaceans belonging to the order Copepoda; abundant members of marine zooplankton communities.

Coriolis effect The deflecting force of the earth's rotation that causes a body of moving water or air to be deflected to the right in the Northern Hemisphere and to the left in the Southern Hemisphere.

Corvid A member of the Covidae family, the jays, crows, and their allies.

Deme A local interbreeding group; any local group of individuals of a given species.

Detritus Organic debris; particles of organic matter derived from the decomposition of plant and animals remains.

Dimorphic Having two forms, as when male and female have different plumages or sizes. Greater Sage-Grouse *(Centrocercus urophasianus)* is an extreme example of a dimorphic species; most raptors show size dimorphism.

Ecotone Boundary or transition zone between adjacent ecological communities.

Edaphic Related to particular soil conditions.

Edge effect An effect exerted by adjoining communities on the population structure within the marginal zone. Generally used to suggest an increase in species richness.

Endemic A type or form that is unique, or restricted, to a place or region. Endemism.

Taxon A taxonomic unit. May refer to an individual population, species, or a group of organisms.

Thermocline A boundary region in the sea between two layers of water of different temperature, within which temperature changes sharply with depth.

Thermoregulation The control of body temperature.

Torpid Dormant; torpidity is like hibernation.

Totipalmate Having fully webbed feet.

Upwelling The movement of cool, nutrient-rich oceanic waters from depth to the surface, usually caused by displacement of surface waters by wind.

Uropygial gland A large gland at the base of a bird's tail that secretes oil used in preening.

Vagrancy A tendency to wander or travel out of the "normal" distributional range of the species. Individuals that occur out of habitat or out of range are considered "vagrants." For example, the Veery *(Catharus fuscescens)* is widely distributed across the northern tier states from Maine to Washington but occurs only as a vagrant in California.

Xeric Extremely dry.

REFERENCES

General

American Ornithologists' Union. 1998. *Check-list of North American birds.* 7th ed. Washington, D.C.: American Ornithologists' Union. Older editions detail subspecies; supplements report taxonomic changes.

American Ornithologists' Union. 2003. Forty-fourth supplement to the checklist of North American birds. *Auk* 120:923–931.

Bailey, F. M. 1902. *Handbook of the birds of western North America.* Boston: Houghton Mifflin.

Bakker, E. S. 1971. *An island called California: An ecological introduction to natural communities.* 2nd ed. Berkeley and Los Angeles: University of California Press.

California Department of Fish and Game and Point Reyes Bird Observatory. 2001. *California bird species of special concern: Draft list and solicitation of input.* www.dfg.ca.gov.

Cooper, D. 2002. *California important bird areas.* Final report. Audubon California. Unpublished.

Coues, E. 1903. *Key to North American birds.* 5th ed. Boston: Page Company Publishers.

Dawson, W. L. 1923. *The birds of California.* 4 vols. San Diego: South Moulton.

Erickson, R. A., and S. N. G. Howell, eds. 2001. *Birds of the Baja California peninsula: Status, distribution, and taxonomy.* Monographs in Field Ornithology, no. 3. Colorado Springs: American Birding Association.

Garrett, K., and J. Dunn. 1981. *Birds of southern California: Status and distribution.* Los Angeles: Los Angeles Audubon Society.

Goudy, C. B., and D. W. Smith. 1994. Ecological units of California: Subsections. Map produced by USDA Forest Service, Pacific

Southwest Region, in cooperation with the Natural Resource Conservation Service. Vallejo, Calif.: USDA Forest Service.

Grenfell, W. E., M. D. Parisi, and D. McGriff. 2003. *Complete list of amphibians, reptiles, birds, and mammals of California.* Sacramento: Wildlife Habitat Relationships System, California Department of Fish and Game in cooperation with the California Interagency Wildlife Task Group.

Grinnell, J., H. C. Bryant, and T. I. Storer. 1918. *The game birds of California.* Berkeley and Los Angeles: University of California Press.

Grinnell, J., and A. H. Miller. 1944. *The distribution and abundance of the birds of California.* Pacific Coast Avifauna, no. 27. Berkeley: Cooper Ornithological Society.

Hall, C. A. 1991. *Natural history of the White-Inyo Range, eastern California.* California Natural History Guides, no. 55. Berkeley and Los Angeles: University of California Press.

Hamilton, R. A., and D. R. Willick. 1996. *The birds of Orange County, California: Status and distribution.* Irvine, Calif.: Sea and Sage Press.

Harris, S. W. 1996. *Northwestern California birds.* 2nd ed. Arcata, Calif.: Humboldt State University Press.

Henshaw, H. W. 1880. Ornithological report from observations and collections made in portions of California, Nevada, and Oregon. In *Appendix L of Annual Report of Geological Survey, west 100th meridian, Chief of engineers report for 1879,* ed. G. M. Wheeler, 282–335.

Hickman, J. C., ed. 1993. *The Jepson manual: Higher plants of California.* Berkeley and Los Angeles: University of California Press.

History of Fresno County, California. 1882. San Francisco: Wallace W. Elliot.

Hoffmann, R. 1927. *Birds of the Pacific states.* Boston: Houghton Mifflin.

Jaeger, E. C., and A. C. Smith. 1966. *Introduction to the natural history of southern California.* California Natural History Guides, no. 13. Berkeley and Los Angeles: University of California Press.

Kemper, J. 2001. *Birding northern California.* A Falcon Guide. Guilford, Conn.: Falcon.

Lehman, P. E. 1994. *The birds of Santa Barbara County, California.* Santa Barbara: Santa Barbara Vertebrate Museum.

Manolis, T. 2003. *Dragonflies and damselflies of California.* California Natural History Guides, no. 72. Berkeley and Los Angeles: University of California Press.

Marshall D. B., M. G. Hunter, and A. L. Contreras, eds. 2003. *Birds of Oregon: A general reference.* Corvallis, Ore.: Oregon State University Press.

McCaskie, G. P. DeBenedictis, R. Erickson, and J. Morlan. 1988. *Birds of northern California: An annotated field list.* Reprinted with supplement. Berkeley: Golden Gate Audubon Society.

Mearns, B., and R. Mearns. 1992. *Audubon to Xantus: The lives of those commemorated in North American bird names.* San Diego: Academic Press.

Muir, J. 1911. *My first summer in the Sierra.* Cambridge, Mass.: Riverside Press.

Munz, P. A. 1968. *Supplement to a California flora.* Berkeley and Los Angeles: University of California Press.

National Audubon Society. 2001. *The Sibley guide to bird life and behavior,* illus. D. A. Sibley, ed. C. Elphick, J. Dunning, and D. Sibley. New York: A. A. Knopf.

Ornduff, R., P. M. Faber, T. Keeler-Wolf. 2003. *Introduction to California plant life.* Rev. ed. California Natural History Guides, no. 69. Berkeley and Los Angeles: University of California Press.

Pavlik, B. M., P. C. Muick, S. Johnson, and M. Popper. 1991. *Oaks of California.* Los Olivos, Calif.: Cachuma Press and California Oak Foundation.

Pyle, P. 1997. *Identification guide to North American birds, Part I: Columbidae to Ploceidae.* Bolinas, Calif.: Slate Creek Press.

Roberson, D. 2002. *Monterey birds: Status and distribution of birds in Monterey County, California.* 2nd ed. Carmel, Calif.: Monterey Audubon Society.

Rosenberg, K. V., R. D. Ohmart, W. C. Hunter and B. W. Anderson. 1991. *Birds of the lower Colorado River valley.* Tucson: University of Arizona Press.

Schoenherr, A. A. 1992. *A natural history of California.* California Natural History Guides, no. 56. Berkeley and Los Angeles: University of California Press.

Shuford, W. D., C. M. Hickey, R. J. Safran, and G. W. Page. 1996. A review of the status of the White-faced Ibis in winter in California. *Western Birds* 27:169–196.

Shuford, W. D., S. A. Laymon, and S. Fitton. 1995. Fifty years after Grinnell and Miller: Organizing for a better future. *Western Birds* 26:205–212.

Sibley, D. A. 2002. *Basic birding.* New York: A. A. Knopf.

Sibley, D. A. 2003. *The Sibley guide to the birds of western North America.* New York: A. A. Knopf.
Small, A. 1974. *The birds of California.* New York: Winchester Press.
Small, A. 1994. *California birds: Their status and distribution.* Vista, Calif.: Ibis Publishing.
Stuart, J. D., and J. O. Sawyer. 2001. *Trees and shrubs of California.* California Natural History Guides, no. 62. Berkeley and Los Angeles: University of California Press.
Unitt, P. 1984. *The birds of San Diego County.* Memoir 13. San Diego: San Diego Society of Natural History.
Welty, J. C., and L. Baptista. 1988. *The life of birds.* 4th ed. New York: W. B. Saunders.
Wolfe, L. M., ed. 1979. *John of the mountains: The unpublished journals of John Muir.* Madison: University of Wisconsin Press.

An Overview of California Birdlife

American Ornithologists' Union. 1957. *Check-list of North American Birds.* 5th ed. Baltimore: Lord Baltimore Press.
American Ornithologists' Union. 1998. *Check-list of North American birds.* 7th ed. Washington, D.C.: American Ornithologists' Union.
Bolger, D. T. 2002. Habitat fragmentation effects on birds in southern California: Contrast to the "top-down" paradigm. In *Effects of habitat fragmentation on birds in westerm landscapes: Contrasts with paradigms from the eastern United States,* ed. T. L. George and D. S. Dobkin. Studies in Avian Biology, no. 25. Allen Press, Lawrence, Kansas: Cooper Ornithological Society.
Carter H. R. , W. R. McIver, and G. J. McChesney 2001. Ashy Storm Petrel *(Oceanodroma homochroa).* In *California bird species of special concern: Draft list and solicitation of input.* California Department of Fish and Game and Point Reyes Bird Observatory. www.dfg.ca.gov/hcpb/bsscindex.
Coues, E. 1874. *Birds of the Northwest: a hand-book of the ornithology of the region drained by the Missouri River and its tributaries.* Washington, D.C.: Government Printing Office.
George, T. L., and D. S. Dobkin. 2002. Introduction. In *Effects of habitat fragmentation on birds in westerm landscapes: Contrasts with paradigms from the eastern United States,* ed. T. L. George and D. S. Dobkin. Studies in Avian Biology, no. 25. Allen Press, Lawrence, Kansas: Cooper Ornithological Society.
Grinnell, J. 1922. The role of the "accidental." *Auk* 39:373–380.

Grinnell, J., H. C. Bryant, and T. I. Storer. 1918. *The game birds of California.* Berkeley and Los Angeles: University of California Press.

Grinnell, J., and A. H. Miller. 1944. *The distribution and abundance of the birds of California.* Pacific Coast Avifauna, no. 27. Berkeley: Cooper Ornithological Society.

Harris, H. 1941. The annals of *Gymnogyps* to 1900. *Condor* 43:1–53.

Henshaw, H. W. 1880. Ornithological report from observations and collections made in portions of California, Nevada, and Oregon. In *Appendix L of Annual Report of Geological Survey, west 100th meridian, Chief of engineers report for 1879,* ed. G. M. Wheeler, 282–335.

Manolis, T. 2003. *Dragonflies and damselflies of California.* California Natural History Guides, no. 72. Berkeley and Los Angeles: University of California Press.

Nuttall, T. 1832-1834. *A Manual of the Ornithology of the U.S. and Canada.* Volume I (Landbirds). Volume II (Waterbirds). Cambridge, Mass.: Hilliard and Brown, Hilliard, Gray and Company.

Rothstein, S. I. 1994. *The Cowbird's invasion of the Far West: History, causes and consequences experienced by host species.* Studies in Avian Biology no. 15. Allen Press, Lawrence, Kansas: Cooper Ornithological Society.

Schoenherr, A. A. 1992. *A natural history of California.* California Natural History Guides, no. 56. Berkeley and Los Angeles: University of California Press.

Shuford, W. D., S. A. Laymon, and S. Fitton. 1995. Fifty years after Grinnell and Miller: Organizing for a better future. *Western Birds* 26:205–212.

Sisk, T. D., and J. Battin. 2002. *Habitat edges and avian ecology: geographic patterns and insights for western landscapes.* Studies in Avian Biology, no. 25. Allen Press, Lawrence, Kansas: Cooper Ornithological Society.

Stattersfield, A. J., M. J. Crosby, A. J. Long, and L. E. Morse. 1998. Endemic bird areas of the world: Priorities for biodiversity conservation. In *California bird species of special concern: Draft list and solicitation of input.* California Department of Fish and Game and Point Reyes Bird Observatory. www.dfg.ca.gov.

Stein, B. A., L. S. Kutner, G. A. Hammerson, L. L. Master, and L. E. Morse. 2000. State of the states: Geographic patterns of diversity, rarity, and endemism. In *Precious heritage: The status of biodiver-*

sity in the United States, ed. B. A. Stein, L. S. Kutner, and J. S. Adams. Chapter 5, 119–157. New York: Oxford University Press.

Tweit, R. C., and D. M. Finch. 1994. Abert's Towhee *(Pipilo aberti).* In *The birds of North America,* no. 111, ed. A. Poole and F. Gill. Philadelphia: Natural Academy of Sciences; and Washington, D.C.: The American Ornithologists' Union.

U.S. Fish and Wildlife Service. 1998. *Recovery plan for Inyo California Towhee (Pipilo crissalis eremophilus).* Portland, Ore.: Region 1, U.S. Fish and Wildlife Service.

Wheelock, I. G. 1910. *Birds of California: An introduction.* 2nd ed. Chicago: A. C. McClurg.

Seabirds and the Marine Environment

Ainley, D. G., and B. J. Boekelheide. 1990. *Seabirds of the Farallon Islands: Ecology, dynamics, and structure of an upwelling-system community.* Palo Alto: Stanford University Press.

Ainley, D. G., and T. J. Lewis, J. 1974. The history of Farallon Island marine bird populations 1843–1972. *Condor* 76:432–46.

Ainley, D. G., R. R. Veit, S. G. Allen, L. B. Spear, and P. Pyle. 1995. Variations in marine bird communities of the California Current, 1986–1994. *California Cooperative Oceanic Fisheries Investigating Reports* 36:72–77.

Alexander, W. B. 1954. *Birds of the ocean.* Rev. ed, 1963. New York: G. P. Putnam's Sons.

Briggs, K. T., W. B. Tyler, D. B. Lewis, and D. R. Carlson. 1987. *Bird communities at sea off California: 1975 to 1983.* Studies in Avian Biology, no. 11. Allen Press, Lawrence, Kansas: Cooper Ornithological Society.

Canright, A. 2002. Rocks and wrecks: Offshore monument. *California Coast and Ocean* 18(2).

Carter, H. R., D. S. Gilmer, J. E. Takekawa, R. W. Lowe, and U. W. Wilson. 2002. *Breeding seabirds in California, Oregon, and Washington.* Dixon, Calif.: National Biological Service, California Pacific Center.

Carter, H. R., G. J., McChesney, D. L. Jacques, C. S. Strong, M. W. Parker, J. E. Takekawa, D. L. Jory, and D. L. Whitworth. 1992. *Breeding populations of seabirds in California 1989–1991.* Vol. 1, *Population estimates.* Unpublished Draft Final Report no. 14-12-001-30456. Camarillo, Calif.: U.S. Minerals Management Service, Pacific OCS Region.

Dawson, W. L. 1923. *The birds of California.* 4 vols. San Diego: South Moulton Company.
DeSante, D., and D. G. Ainley. 1980. *The avifauna of the South Farallon Islands. California.* Studies in Avian Biology, no. 4. Allen Press, Lawrence, Kansas: Cooper Ornithological Society.
Haley, D. 1984. *Seabirds of the eastern North Pacific and Arctic waters.* Seattle: Pacific Search Press.
Hoffmann, R. 1927. *Birds of the Pacific states.* Boston: Houghton Mifflin Company.
Oedekoven, C. S., D. G. Ainley, and L. B. Spear. 2001. Variable responses to change in marine climate: California Current, 1985–1994. *Marine Ecology Progress Series* 212:265–281.
Schoenherr, A. A., C. R. Feldmeth, and M. J. Emerson. 1999. *Natural history of the islands of California.* Berkeley and Los Angeles: University of California Press.
Stallcup, R. 1990. *Ocean birds of the nearshore Pacific: A guide for the sea-going naturalist.* Stinson Beach, Calif.: Point Reyes Bird Observatory.
Stewart, K. 1997. Seabirds smell sea-smells from the seascape. *Davis Enterprise* (California), 30 March.
Veit, R. R., P. Pyle, and J. A. McGowan. 1996. Ocean warming and long term change in pelagic bird abundance within the California current system. *Marine Ecology Progress Series* 139:11–18.
Wheelock, I. G. 1910. *Birds of California: an introduction.* Chicago: A. C. McClurg & Co.

Birds of the Shoreline

Accurso, L. M. 1992. Distribution and abundance of wintering waterfowl on San Francisco Bay: 1988–1990. M.S. thesis, Humboldt State University.
Bildstein, K. L., G. T. Bancroft, P. J. Dungan, D. H. Gordon, R. M. Erwin, E. Nol, L. X. Payne, and S. E. Steiner. 1991. Approaches to conservation of coastal wetlands in the Western Hemisphere. *Wilson Bulletin* 103:218–254.
Bruce, C., K. Miller, G. Page, L. Stenzel, and W. White. 1994. Western Snowy Plover. In *Life on the Edge.* Vol. 1, *Wildlife.* Santa Cruz, Calif.: Biosystems Books.
Canright, A. 2002. Rocks and Wrecks: Offshore Monument. *California Coast and Ocean* 18(2).
Chase, M. K., W. B. Kristan III, A. J. Lynam, M. V. Price, and J. T. Rotenberry. 2000. Single species as indicators of species richness

and composition in California coastal sage scrub birds and small mammals. *Conservation Biology* 14:474–487.

Colwell, M. A., and S. L. Landrum. 1993. Nonrandom shorebird distribution and fine-scale variation in prey abundance. *Condor* 95:94–103.

Connors, P. G., J. P. Myers, C. S. Connors, and F. A. Pitelka. 1981. Interhabitat movements by Sanderlings in relation to foraging profitability and the tidal cycle. *Auk* 98:49–64.

Cramp, S., ed. 1983. *The handbook of the birds of Europe, the Middle East and North Africa: Birds of the western Palearctic.* Vol. 3, *Waders to gulls.* Oxford: Oxford University Press.

Garrett, K., and J. Dunn. 1981. *Birds of southern California: Status and distribution.* Los Angeles: Los Angeles Audubon Society.

Goals Project. 1999. *Baylands ecosystem habitat goals.* A report of habitat recommendations prepared by the San Francisco Bay Area Wetlands Ecosystem Goals Project. San Francisco: U.S. Environmental Protection Agency; Oakland: San Francisco Bay Regional Water Quality Control Board.

Grinnell, J. H., C. Bryant, and T. I. Storer. 1918. *The game birds of California.* Berkeley and Los Angeles: University of California Press.

Harrington, B., and E. Perry. 1995. Important shorebird staging sites meeting Western Hemisphere Shorebird Reserve Network criteria in the United States. Washington, D.C.: U.S. Fish and Wildlife Service report.

Howell, S. N. G., and T. Gardali. 2003. Phenology, sex ratios, and population trends of *Selasphorus* hummingbirds in central Coastal California. *Journal of Field Ornithology* 74(1): 17–25.

Kelly, J., H. M. Pratt, and P. L. Greene. The distribution, reproductive success, and habitat characteristics of heron and egret breeding colonies in the San Francisco Bay Area. *Colonial Waterbirds* 16(1): 18–27.

Kelly, J., and S. Tappen. 1998. Distribution, abundance, and implications for conservation of winter waterbirds on Tomales Bay, California. *Western Birds* 29:103–120.

Matthiessen, P. 1967. *The windbirds.* New York: Viking Press.

Matthiessen, P., R. S. Palmer, and G. Stout. 1967. *The shorebirds of North America.* New York: Viking Press.

Myers, J. P. 1980. Sanderling *Calidris alba* at Bodega Bay: Facts, inferences and shameless speculations. *Wader Study Group Bulletin* 30:26–32.

Nur, N., S. Zack, J. Evens, and T. Gardali. 1997. Tidal marsh birds of the San Francisco Bay Region: Status, distribution, and conservation of five category 2 taxa. Final Report to U.S. Geological Survey, Biological Resources Division. Washington, D.C.: U.S. Geological Survey.

Page, G., L. E. Stenzel, and J. E. Kjelmyr. 1999. Overview of shorebird abundance and distribution in wetlands of the Pacific Coast of the contiguous United States. *Condor* 101:461–471.

Page, G., L. E. Stenzel, and C. M. Wolfe. 1979. *Aspects of occurrence of shorebirds on a central California estuary.* Studies in Avian Biology, No. 2. Allen Press, Lawrence, Kansas: Cooper Ornithological Society.

Page, G. F., C. Bidstrip, R. J. Ramer, and L. E. Stenzel. 1986. Distribution of wintering Snowy Plovers in California and adjacent states. *Western Birds* 17:145–170.

Page, G. W., and L. E. Stenzel, eds. 1981. The breeding status of the Snowy Plover in California. *Western Birds* 12:1–40.

Rubinoff, D. 2001. Evaluating the California Gnatcatcher as an umbrella species for conservation of southern California coastal sage scrub. *Conservation Biology* 15:1374–1383.

Shuford, W. D., G. W. Page, J. G. Evens, and L. E. Stenzel. 1989. Seasonal abundance of waterbirds at Point Reyes: A coastal California perspective. *Western Birds* 20:137–265.

Shuford, W. D., G. W. Page, and C. M. Hickey. 1995. Distribution and abundance of Snowy Plover wintering in the interior of California and adjacent states. *Western Birds* 26:82–98.

Stenzel, L., H. R. Huber, and G. W. Page. 1976. Feeding behavior and diet of the Long-billed Curlew and the Willet. *Wilson Bulletin* 88(2): 314–332.

Stenzel, L., G. W. Page, and J. Young. 1983. *The tropic relationships between shorebirds and their prey on Bolinas Lagoon.* Stinson Beach, Calif.: Point Reyes Bird Observatory.

Stenzel, L. E., C. M. Hickey, J. E. Kjelmyr, and G. W. Page. 2002. Abundance and distribution of shorebirds in the San Francisco Bay Area. *Western Birds* 33:69–98.

Stenzel, L. E., J. C. Warriner, J. S. Warriner, K. S. Wilson, F. C. Bidstrip, and G. W. Page. 1994. Long-distance breeding dispersal of Snowy Plovers in western North America. *Journal of Animal Ecology* 63:887–902.

Takakawa, J. Y., G. W. Page, J. M. Alexander, and D. R. Becker. 2000. Waterfowl and shorebirds of the San Francisco Bay Estuary. In *Baylands ecosystem species and community profiles: Life histories and environmental requirements of key plants, fish and wildlife*, ed. P. R. Olofson. Prepared by the San Francisco Bay Area Wetlands Ecosystem Goals Project. Oakland: San Francisco Bay Regional Water Quality Control Board.

Tuttle, D. C., and R. Stein. 1997. Snowy Plover nesting on Eel River gravel bars, Humboldt County. *Western Birds* 28:174–176.

U.S. Fish and Wildlife Service. 1999. Waterfowl survey data from San Francisco Bay. In *Restoring the estuary.* Portland, Ore.: U.S. Fish and Wildlife Service, Region 1.

U.S. Fish and Wildlife Service. 1998–2000. Aerial waterfowl surveys of San Francisco Bay. Unpublished data.

Warnock, N., G. W. Page, and L. Stenzel. 1995. Non-migratory movements of Dunlins on their California wintering grounds. *Wilson Bulletin* 107:131–139.

Warnock, N., G. W. Page, T. D. Ruhlen, N. Nur, J. Y. Takekawa, and J. T. Hanson. 2002. Management and conservation of the San Francisco Bay salt ponds: Effects on pond salinity, area, tide, and season on Pacific Flyway waterbirds. *Waterbirds* Special Publication 2:79–92.

Warnock, S. E., and J. Y. Takekawa. 1995. Habitat preferences of wintering shorebirds in a temporally changing environment: Western Sandpipers in the San Francisco Bay Estuary. *Auk* 112:920–930.

Birds of the Coast Ranges

Atwood, J. L. 1988. *Speciation and geographic variation in Black-tailed Gnatcatchers.* Ornithological Monographs 42. Fayetteville, Ark.: American Ornithologists' Union.

Binford, L. C. 1979. Fall migration of diurnal raptors at Pt. Diablo, California. *Western Birds* 10:1–16.

Block, W. M., M. L. Morrison, and J. Verner. 1990. Wildlife and oak-woodland interdependency. *Fremontia* 18(3):72–76.

DeSante, D., and D. G. Ainley. *The avifauna of the South Farallon Islands, California.* Studies in Avian Biology, no. 4. Allen Press, Lawrence, Kansas: Cooper Ornithological Society.

Evens, J. 2001. The San Francisco Common Yellowthroat *(Geothlypis trichas sinuosa)*. In *California bird species of special concern: Draft list and solicitation of input.* California Department of Fish and Game and Point Reyes Bird Observatory. www.dfg.ca.gov.

Gardali, T., and A. Jaramillo. 2001. Further evidence for a population decline in the western Warbling Vireo. *Western Birds* 32:173–176.

Grinnell, J. Up-hill planters. *Condor* 38(2): 80–82.
Hall, L. S., A. M. Fish, and M. L. Morrison. 1992. The influence of weather on hawk movements in coastal northern California. *Wilson Bulletin* 104:447–461.
Hickman, J. C., ed. 1993. *The Jepson manual: Higher plants of California.* Berkeley and Los Angeles: University of California Press.
Kelly, E. G., E. D. Forsman, and R. G. Anthony. 2003. Are Barred Owls displacing Spotted Owls? *Condor* 105:45–53.
Koenig, W. 1990. Oaks, acorns, and the Acorn Woodpecker. *Fremontia* 18(3): 77–79.
Morro Coast Audubon Society. 1995. *The birds of San Luis Obispo County, California.* Morro Bay, Calif.: Morro Coast Audubon Society.
Rotenberry, J. T., and T. A. Scott. 1998. Biology of the California Gnatcatcher: Filling in the gaps. *Western Birds* 29:237–241.
Shuford, W. D. 1993. *The Marin County breeding bird atlas: A distributional and natural history of coastal California birds.* California Avifauna Series 1. Bolinas, Calif.: Bushtit Books.
Shuford, W. D., and I. C. Timossi. 1989. *Plant communities of Marin County.* Special Publication no. 10. Sacramento: California Native Plant Society.
Sterling, J., and P. W. C. Patton. 1996. Breeding distribution of Vaux's Swift in California. *Western Birds* 27:30–40.
Zink, R. M., G. F. Barrowclough, J. L. Atwood, and R. C. Blackwell-Rago. 2000. Genetics, taxonomy, and conservation of the threatened California Gnatcatcher. *Conservation Biology* 14:1394–1405.

Birds of the Central Valley and Delta

American Ornithologists' Union. 1998. *Check-list of North American birds.* 7th ed. Washington, D.C.: American Ornithologists' Union.
Bay Institute. 1998. *From the Sierra to the sea: The ecological history of the San Francisco Bay-Delta watershed.* San Rafael, Calif.: The Bay Institute of San Francisco.
Cole, L. 2003. *Birding in Kings County.* Site guide reprinted from the Central Valley Bird Club Bulletin. www.cvbirds.org.
Fresno Audubon Society. 2002. *Birds of Fresno and Madera Counties.* www.fresnoaudubon.org.
Grinnell, J., H. C. Bryant, and T. I. Storer. 1918. *The game birds of California.* Berkeley and Los Angeles: University of California Press.

Grinnell, J., and A. H. Miller. 1944. *The distribution and abundance of the birds of California.* Pacific Coast Avifauna, no. 27. Berkeley: Cooper Ornithological Society.

Humple, D., and G. Geupel. 2002. Autumn populations of birds in riparian habitat of California's Central Valley. *Western Birds* 33:34–50.

Kirk, A. 1994. Vanished lake, vanished landscape. In *Life on the edge: A guide to California's endangered natural resources.* Santa Cruz, Calif.: Biosystems Books.

Knopf, F. L. 1988. Conservation of steppe birds in North America. In *Ecology and conservation of grassland birds,* ed. P. D. Goriup. Technical Publication 7. Cambridge, U.K.: International Council for Bird Preservation.

Laymon, S., and M. D. Halterman. 1987. Can the western subspecies of the Yellow-billed Cuckoo be saved from extinction? *Western Birds* 18:19–25.

Littlefield, C. D. 2002. Winter foraging habitat of Greater Sandhill Cranes in northern California. *Western Birds* 33:51–60.

Manolis, T. 2003. *Dragonflies and damselflies of California.* California Natural History Guides, no. 72. Berkeley and Los Angeles: University of California Press.

Manolis, T., and G. V. Tangren. 1975. Shorebirds of the Sacramento Valley, California. *Western Birds* 6:45–54.

Nordoff, C. 1873. California: For health, pleasure, and residence. Reprinted in Dawson, W. L. 1923. *The Birds of California.* 4 vols. San Diego: South Moulton Company.

Pogson, T., and S. M. Lindstedt. 1991. Distribution and abundance of large Sandhill Cranes, *Grus canadensis,* wintering in California's Central Valley. *Condor* 93:266–278.

Reeve, H. 2003. *Birding the Modesto sewage ponds.* Site guide reprinted from the Central Valley Bird Club Bulletin. www.cvbirds.org

Shuford, W. D., G. W. Page, and C. M. Hickey. 1995. Distribution and abundance of Snowy Plover wintering in the interior of California and adjacent states. *Western Birds* 26:82–98.

Shuford, W. D., G. W. Page, and J. E. Kjelmyr. 1998. Patterns and dynamics of shorebird use of California's Central Valley. *Condor* 100:227–244.

Sivera, J. G. 1998. Avian uses of vernal pools and implications for conservation practice. In *Ecology, conservation, and management of vernal pool ecosystems,* ed. C. W. Witham, E. T. Bauder, D. Belk,

W. R. Ferren, Jr., and R. Ornduff. Sacramento: California Native Plant Society.
Thompson, K. 1961. Riparian forests of the Sacramento Valley. *Annals of the Association of American Geographers* 51(3): 294–315.
Thompson, K. 1980. Riparian forests of the Sacramento Valley, California. In *Riparian forests of California,* ed. A. Sands. Davis, Calif.: Division of Agricultural Science, University of California, Davis.
U.S. Fish and Wildlife Service. 1990. Central Valley habitat joint venture implementation plan. A component of the North American Waterfowl Management Plan. Sacramento: U.S. Fish and Wildlife Service.
Williamson, R. S. 1853. *Corps and topographical engineers report of explorations in California for railroad routes to connect with the routes near the 35th and 32nd parallels of north latitude.* Washington, D.C.: War Department.
Wunder, M. B., and F. L. Knopf. 2003. The Imperial Valley of California is critical to wintering Mountain Plovers. *Journal of Field Ornithology* 74(1): 74–80.

Birds of Mountains and Foothills

Adkisson, C. S. 1999. Pine Grosbeak *(Pinicola enucleator).* In *The birds of North America,* no. 456, ed. A. Poole and F. Gill. Philadelphia: The Birds of North America.
American Ornithologists' Union. 1983. *Check-list of North American birds.* 6th ed. Washington, D.C.: American Ornithologists' Union.
Barnes, B., and B. Steele. 2003. The Kern River Valley and southern Sierra of California. *Birding* 35(2): 156–166.
Beedy, E. C., and S. L. Granholm. 1985. *Discovering Sierra birds.* Three Rivers, Calif.: Yosemite Natural History Association and Sequoia Natural History Association. 229 pp.
Bland, J. 2001. Mount Pinos Blue Grouse. In *California bird species of special concern: Draft list and solicitation of input.* California Department of Fish and Game and Point Reyes Bird Observatory. www.dfg.ca.gov.
Bull, E., and J. R. Duncan. 1993. Great Gray Owl *(Strix nebulosa).* In *The birds of North America,* no. 41, eds. A. Poole and F. Gill. Philadelphia: Academy of Natural Sciences; Washington, D.C.: American Ornithologists' Union.
Dawson, W. L. 1923. *The birds of California.* 4 vols. San Diego: South Moulton Company.

Dunn, J. 2002. Looking for woodpeckers? Winging it. *American Birding Association Newsletter,* May 2002.
Gabrielson, I. N., and S. G. Jewitt. 1940. *Birds of Oregon.* Corvallis: Oregon State College.
Gaines, D. 1977. *Birds of the Yosemite Sierra: A distributional survey.* Oakland, Calif.: California Syllabus Press.
Gould, G. I. 2001. California Spotted Owl *(Strix occidentalis occidentalis).* In *California bird species of special concern: Draft list and solicitation of input.* California Department of Fish and Game and Point Reyes Bird Observatory. www.dfg.ca.gov.
Grinnell, J., H. C. Bryant, and T. I. Storer. 1918. *The game birds of California.* Berkeley and Los Angeles: University of California Press.
Johnson, N. K., and C. B. Johnson. 1985. Speciation in sapsuckers *(Sphyrapicus):* Sympatry, hybridization, and mate preference in *S. ruber daggettii* and *S. nuchalis. Auk* 102:1–15.
Lentz, J. E. 1987. Breeding birds of four isolated mountains in southern California. *Western Birds* 24:201–234.
McCallum, D. A. 1994. Flammulated Owl *(Otus flammeolus).* In *The Birds of North America,* no. 93, ed. A. Poole and F. Gill. Philadelphia: The Birds of North America.
McCallum, D. A., R. Grundel, and D. L. Dahlsten. 1999. Mountain Chickadee *(Poecile gambeli).* In *The Birds of North America,* no. 453, ed. A. Poole and F. Gill. Philadelphia: The Birds of North America.
Miller, A. H., and R. C. Stebbins. 1964. *The lives of desert animals in Joshua Tree National Monument.* Berkeley and Los Angeles: University of California Press.
Rowe, S. P. and T. Gallion. 1996. Fall migration of Turkey Vultures and raptors through the southern Sierra Nevada, California. *Western Birds* 27:48–53.
Shuford, W. D., and S. D. Fitton. 1998. Status of owls in the Glass Mountain Region, Mono County, California. *Western Birds* 29:1–20.
Storer, T. I., and R. L. Usinger. 1963. *Sierra Nevada natural history.* Berkeley and Los Angeles: University of California Press.
Winter, J. 1974. The distribution of the Flammulated Owl in California. *Western Birds* 5:25–44.

Birds of the Great Basin

Able, K. P. 2000. Sage Grouse futures. *Birding* 32:306–316.
Behle, W. H. 1978. Avian biogeography of the Great Basin and inter-

mountain region. In *Intermountain biogeography: A symposium.* Great Basin Naturalist Memoirs 2. Baker, Nev.: National Park Service, Great Basin National Park.

Braun, C. E., and B. T. Wallestad. 1977. Guidelines for maintenance of sage grouse habitats. *Wildlife Society Bulletin* 5:99–106.

Bureau of Land Management, U.S. Fish and Wildlife Service, U.S. Forest Service, Oregon Department of Fish and Wildlife, Oregon Department of State Lands. 2000. Greater Sage-Grouse and sagebrush-steppe ecosystems: Management guidelines. Portland, Ore.: Bureau of Land Management.

DeSante, D. F., and T. L. George. 1994. Population trends in the land birds of western North America. In *A century of avifaunal change in western North America,* ed. J. R. Jehl Jr., and N. K. Johnson. Studies in Avian Biology, no. 15. Cooper Ornithological Society.

Gates, R. J. 1985. Observations of the formation of a sage grouse lek. *Wilson Bulletin* 97:219–221.

Grinnell, J., and A. H. Miller. 1944. *The distribution and abundance of the birds of California.* Pacific Coast Avifauna, no. 27. Berkeley: Cooper Ornithological Society.

Herzog, S. K. 1996. Wintering Swainson's Hawks in California's Sacramento–San Joaquin River Delta. *Condor* 98:876–879.

Hickman, J. C., ed. 1993. *The Jepson manual: Higher plants of California.* Berkeley and Los Angeles: University of California Press.

Jehl, J. R., Jr. 1994. Changes in saline and alkaline lake avifaunas in western North America in the past 150 years. In *A century of avifaunal change in western North America,* ed. J. R. Jehl Jr., and N. K. Johnson. Studies in Avian Biology, no. 15. Cooper Ornithological Society.

Jehl, J. R., and D. R. Jehl. 1981. A North American record of the Asiatic marbled murrelet *(Brachhyramphus marmoratus perdix). American Birds* 36:91–92.

Johnson, N. K. 1975. Controls of number of bird species on montane islands in the Great Basin. *Evolution* 29:545–567.

Johnson, N. K. 1978. Patterns of avian geography and speciation in the Great Basin. In *Intermountain biogeography: A symposium,* ed. K. T. Harper and J. L. Reveal. Great Basin Naturalist Memoirs 2. Baker, Nev.: National Park Service, Great Basin National Park.

Knick, S. T., and J. T. Rotenberry. 1995. Landscape characteristics of fragmented shrub steppe habitats and breeding passerine birds. *Conservation Biology* 9:1059–1071.

National Science Foundation. 1987. *The Mono Basin ecosystem:*

Effects of changing lake level. National Research Council: Mono Basin Ecosystem Study Committee. Washington, D.C.: National Academy Press.

Page, G. W., L. E. Stenzel, D. W. Winkler, and C. W. Swarth. 1983. Spacing out at Mono Lake: Breeding success, nest density and predation in the Snowy Plover. *Auk* 100:13–24.

Paige, C., and S. A. Ritter. 1999. *Birds in a sagebrush sea: Managing sagebrush habitats for bird communities.* Boise, Idaho: Partners in Flight, Western Working Group.

Peterjohn, B. G., J. R. Sauer, and W. A. Link. 1994. The 1992–1993 summary of the North American Breeding Bird Survey. *Bird Populations* 2:246–261.

Rich, T. 2001. Birds and the war on weeds. *Birding* 33(3): 241–248.

Rich T., and B. Altman. 2000. Under the Sage Grouse umbrella. *Bird Conservation* 14:10.

Rubinoff, D. 2001. Evaluating the California Gnatcatcher as an umbrella species for conservation of southern California coastal sage shrub. *Conservation Biology* 15:1374–1383.

Ryser, F., Jr. 1985. *Birds of the Great Basin: A natural history.* Reno, Nev.: University of Nevada Press.

Schroeder, M. A., J. R. Young, and C. E. Braun. 1999. Sage Grouse *(Centrocercus urophasianus).* In *The Birds of North America,* no. 425, ed. A. Poole and F. Gill. Philadelphia: The Birds of North America.

Schulz, T. T., and W. C. Leininger. 1991. Nongame wildlife communities in grazed and ungrazed montane riparian sites. *Great Basin Naturalist* 51:286–292.

Shuford, W. D., and S. D. Fitton. 1998. Status of owls in the Glass Mountain region, Mono County, California. *Western Birds* 29:1–20.

Shuford, W. D., and T. P. Ryan. 2000. Nesting populations of California and Ring-billed Gulls in California: Recent surveys and historical status. *Western Birds* 31:133–164.

Strauss, E. A., C. A. Ribic, and W. D. Shuford. 2002. Abundance and distribution of migratory shorebirds at Mono Lake, California. *Western Birds* 33:222–240.

Stuart, J. D., and J. O. Sawyer. 2001. *Trees and shrubs of California.* California Natural History Guides, no. 62. Berkeley and Los Angeles: University of California Press.

Tueller, P. T., R. J. Tausch, and V. Bostick. 1991. Species and plant community distribution in a Mojave Great Basin Desert transition. *Vegetation* 92:133–150.

U.S. Fish and Wildlife Service. 1997. *Klamath/Central Pacific Coast ecoregion restoration strategy,* vol. 4. Portland, Ore.: U.S. Fish and Wildlife Service, Region 1.

Wakkinen, W. L., K. P. Reese, and J. W. Connelly. 1992. Sage grouse nest locations in relation to leks. *Journal of Wildlife Management* 56:381–383.

Wiens, J. A., and J. T. Rotenberry. 1981. Habitat associations and community structure of shrub-steppe environments. *Ecological Monographs* 51:21–41.

The Deserts' Birds

American Ornithologists' Union. 1998. *Check-list of North American birds.* 7th ed. Washington, D.C.: American Ornithologists' Union.

Atwood, J. L. 1988. *Speciation and geographic variation in the black-tailed gnatcatchers.* Ornithological Monographs 42.

Brown, D. E., J. C. Hagelin, M. Taylor, and J. Galloway. 1998. Gambel's Quail *(Callipepla gambelii).* In *The Birds of North America,* no. 321. Philadelphia: The Birds of North America.

Cannings, R. J., and T. Angell. 2001. Western Screech-Owl *(Otus kennicottii).* In *The Birds of North America,* no. 597, ed. A. Poole and F. Gill. Philadelphia: Natural Academy of Sciences; Washington, D.C.: The American Ornithologists' Union.

Cardiff, S. W., and D. L. Dittmann. 2000. Brown-crested Flycatcher *(Myiarchus tyrannulus).* In *The Birds of North America,* no. 496. ed. A. Poole and F. Gill. Philadelphia: Natural Academy of Sciences; and Washington, D.C.: The American Ornithologists' Union.

Cardiff, S. W., and J. V. Remsen. 1981. Breeding avifaunas of the New York Mountains and Kingston Range: Islands of conifers in the Mohave Desert of California. *Western Birds* 12:73–86.

Dawson, W. L. 1923. *The birds of California.* 4 vols. San Diego: South Moulton.

DeSante, D. F., and E. D. Ruhan. 1995. *A census of burrowing owls in California, 1991–1993.* Point Reyes Station, Calif.: Institute for Bird Populations.

Farquhar, C. C., and K. L. Ritchie. 2002. Black-tailed Gnatcatcher *(Polioptila melanura).* In *The Birds of North America,* no. 690, ed. A. Poole and F. Gill. Philadelphia: Natural Academy of Sciences; and Washington, D.C.: The American Ornithologists' Union.

Franzreb, K. E. 1978. Breeding bird densities, species composition, and bird species diversity at Algodones dunes. *Western Birds* 9:9–20.

Grinnell, J. 1908. Birds of a voyage on the Salton Sea. *Condor* 10: 185–191.
Hoffmann, R. 1927. *Birds of the Pacific states.* Boston: Houghton Mifflin.
Hunting, K. 2003. *Mountain Plover.* California bird species of special concern. www.dfg.ca.gov/hcpb/bsscindex.html.
Jaeger, E. 1933. The California deserts. Palo Alto: Stanford University Press. 4th edition 1965.
Jaeger, E. 1948. Does the Poor-will hibernate? *Condor* 50:45–46.
Jaeger, E. C. 1961. *Desert wildlife.* Stanford, Calif.: Stanford University Press.
Johnson, N. K. *Pioneering and natural expansion of breeding distributions in western North American birds.* Studies in Avian Biology, no. 15. Cooper Ornithological Society.
Johnson, N. K., and K. L. Garrett. 1974. Interior bird species expand their breeding ranges into southern California. *Western Birds* 5:45–56.
Johnson, S. 2000. Disease strikes again at Salton Sea. *Endangered Species Bulletin* 25(5): 6–7.
Massey, B. W., and R. Zembel. 2002. *Guide to birds of the Salton Sea.* Tuscon: Arizona–Sonora Desert Museum.
Mathias, M. E. 1978. The California desert. *Fremontia* 6(3): 3–6.
Miller, A. 1951. An analysis of the distribution of the birds of California. *University of California Publications in Zoology* 50:531–644.
Miller, A. H., and R. C. Stebbins. 1964. *The lives of desert animals in Joshua Tree National Monument.* Berkeley and Los Angeles: University of California Press.
Molina, K. C. 1996. Population status and breeding biology of Black Skimmers at the Salton Sea, California. *Western Birds* 27: 143–158.
Omland, K. 2001. Thrashing out species limits in the southwest. *Birding* 33(4): 320–327.
Patten, M., G. McCaskie, and P. Unitt. 2003. *Birds of the Salton Sea: Status, biogeography, and ecology.* Berkeley and Los Angeles: University of California Press.
Patten, M. A., E. Mellick, H. Gomez de Silva, and T. E. Wurster. 2001. Status and taxonomy of the Colorado Desert avifauna of Baja California. In *Birds of the Baja California Peninsula: Status, distribution, and taxonomy.* Monographs in Field Ornithology, no. 3. Colorado Springs, Co.: American Birding Association.

Pemberton, J. R. 1927. The American Gull-billed Tern breeding in California. *Condor* 29:253–258.

Rosenberg, K. V., R. D. Ohmart, W. C. Hunter, and B. W. Anderson. 1991. *Birds of the Lower Colorado Valley.* Tuscon: University of Arizona Press.

Schoenherr, A. A. 1992. *A natural history of California.* California Natural History Guides, no. 56. Berkeley and Los Angeles: University of California Press.

Shuford, W. D., and K. C. Molina, eds. 2004. *Ecology and conservation of birds of the Salton Sink: An endangered ecosystem.* Studies in Avian Biology, no. 27. Allen Press, Lawrence, Kansas: Cooper Ornithological Society.

Shuford, W. D., N. Warnock, and K. C. Molina. 1999. *The avifauna of the Salton Sea: A synthesis.* Stinson Beach, Calif.: Point Reyes Bird Observatory.

Shuford, W. D., N. Warnock, K. C. Molina, and K. K. Strumm. 2002. The Salton Sea as critical habitat to migratory and resident waterbirds. *Hydrobiologia* 473:255–274.

Skutch, A. 1996. *Orioles, blackbirds, and their kin: A natural history.* Tucson: University of Arizona Press.

Tweit, R. C. 1996. Curve-billed Thrasher *(Toxostoma curvirostre).* In *The Birds of North America,* no. 235, ed. A. Poole and F. Gill. Philadelphia: Natural Academy of Sciences; Washington, D.C.: The American Ornithologists' Union.

Unitt, P. 1981. Birds of Hot Springs Mountain, San Diego County, California. *Western Birds* 12:125–135.

U.S. Fish and Wildlife Service. 1998. *Recovery plan for Inyo California Towhee* (Pipilo crissalis eremophilus). Portland, Ore.: U.S. Fish and Wildlife Service, Region 1.

Weathers, W. W. 1983. *Birds of southern California's Deep Canyon.* Berkeley and Los Angeles: University of California Press.

Webster, M. D. 1999. Verdin *(Auriparus flaviceps).* In *The Birds of North America,* No. 470, ed. A. Poole and F. Gill. Philadelphia: Natural Academy of Sciences and Washington, D.C.: The American Ornithologists' Union.

Zink, R. M., G. F. Barrowclough, J. L. Atwood, and R. C. Blackwell-Rago. 2000. Genetics, taxonomy, and conservation of the threatened California Gnatcatcher. *Conservation Biology* 14:1394–1405.

ADDITIONAL CAPTIONS

PAGES XII–1 The Channel Islands; looking towards Santa Cruz Island from Anacapa.

PAGES 38–39 Nesting Common Murres and Brandt's Cormorants coexist in colonies on South Farallon Island.

PAGES 78–79 Meadows above a rocky California shoreline.

PAGES 114–115 Oak woodland north of San Francisco.

PAGES 152–153 The Sutter Buttes are a volcanic intrusion into the floor of the Central Valley. Their heights provide a haven for raptors that prey on the multitudes of wildfowl passing through the valley.

PAGES 186–187 The rugged landscape of Yosemite is typical austere montane habitat.

PAGES 226–227 The Owens Valley is an extension of the Great Basin flanked by the Sierras to the west and the White Mountains to the east.

PAGES 260–261 The Sonoran Desert is a landscape dominated by ocotillo and a variety of cactus species.

INDEX OF BIRDS

GENERAL INDEX

ABOUT THE AUTHORS

Jules Evens has been a field biologist in California for 30 years; his primary focus is on the ecology of wetland birds. He is a research associate with Avocet Research, the Point Reyes Bird Observatory, and Audubon Canyon Ranch. He is also the author of *The Natural History of the Point Reyes Peninsula.*

Ian Tait is a wildlife and nature photographer whose primary interest over the last 25 years has been birds. His work has appeared in scientific and popular journals as well as museum exhibits. He is an Associate of the Royal Photographic Society.

Series Design: Barbara Jellow
Design Enhancements: Beth Hansen
Design Development: Jane Tenenbaum
Cartographer: Bill Nelson
Composition: Jane Rundell
Text: 9/10.5 Minion
Display: ITC Franklin Gothic Book and Demi
Printer and binder: Everbest Printing Company